PLASMA PROCESSING OF MATERIALS

PLASMA PROCESSING OF MATERIALS

Edited by

H.J. Oskam

Center of Plasma Chemistry
Institute of Technology
University of Minnesota
Minneapolis, Minnesota

np

NOYES PUBLICATIONS
Park Ridge, New Jersey, USA

Library of Congress Catalog Card Number 84-14806
ISBN: 0-8155-1003-9
Printed in the United States

Published in the United States of America by
Noyes Publications
Mill Road, Park Ridge, New Jersey 07656

10 9 8 7 6 5 4 3 2 1

Library of Congress Cataloging in Publication Data
Main entry under title:

Plasma processing of materials.

Bibliography: p.
Includes index.
1. Plasma engineering. I. Oskam, H.J. (Hendrik Jan)
TA2020.P53 1984 621.042 84-14806
ISBN 0-8155-1003-9

Foreword

This book presents detailed studies relating to plasma chemistry and plasma processing of materials. Experimental systems are described for plasma etching and plasma deposition of thin films.

Basic parameters and mechanisms for volume synthesis of materials are also investigated. These syntheses involve the use of radiofrequency (rf) discharges to effect plasma processing on a scale of grams.

Plasma chemistry deals with the chemical and physical reactions which occur between particles (ions, electrons, and neutral species) produced by an electrical discharge in a partially ionized gas, called the "plasma," and/or between these particles and material surfaces in contact with them. Reactions in a plasma are induced by electron impact resulting from the electrical discharge. The degree of plasma formation is controlled by regulation of pressure and applied power, lower pressure and/or higher power generally yielding more energetic plasmas.

The information in the book is from *Study of Plasma Chemistry and Plasma Processing. December 15, 1979 to June 15, 1983. Final Technical Report,* prepared by H.J. Oskam, University of Minnesota Center of Plasma Chemistry, for the U.S. Department of the Navy Office of Naval Research, 1983.

The table of contents is organized in such a way as to serve as a subject index and provides easy access to the information contained in the book.

Advanced composition and production methods developed by Noyes Publications are employed to bring this durably bound book to you in a minimum of time. Special techniques are used to close the gap between "manuscript" and "completed book." In order to keep the price of the book to a reasonable level, it has been partially reproduced by photo-offset directly from the original report and the cost saving passed on to the reader. Due to this method of publishing, certain portions of the book may be less legible than desired.

NOTICE

The materials in this book were prepared as accounts of work sponsored by the U.S. Department of the Navy. Publication does not signify that the contents necessarily reflect the views and policies of the contracting agency or the publisher, nor does mention of trade names or commercial products constitute endorsement or recommendation for use.

Contents and Subject Index

I. Introduction

The studies described in this final technical report were funded under a three year contract with the Office of Naval Research (Contract #N00014-80-C-0244, NR SRO-016). The studies were performed during the period December 15, 1979 through June 15, 1983, where the last one half year was a no-cost extension.

The purpose of the studies was to obtain a better understanding of the physical and chemical concepts of importance during the low pressure plasma processing of materials. Plasma chemistry deals with the occurrence of chemical and physical reactions between particles produced in a partially ionized gas (ions, electrons, and neutral particles) and/or between these particles and material surfaces in contact with the ionized gas.

The studies performed can be divided into three different but related areas. The studies were started using two existing experimental systems. During the contract period two new experimental systems were designed and constructed. Professor Oskam had the primary responsibility for this effort, while Dr. Ernie (post doctoral fellow) supplied the details of the design and supervised the construction. Section II describes these two new experimental systems, which make it possible to study basic phenomena related to plasma etching and plasma deposition of thin films.

The second part of the research effort concerns the plasma processing of polymer films. Professor Evans had the primary responsibilities for these studies. The results are described in detail in section III. The third part of the studies relates to plasma volume synthesis of materials for which Dr. Miller had the primary responsibility. These studies are discussed in section IV.

Appendix I lists the publications in scientific journals and presentations at conferences during the contract period. The personnel (staff and students) involved in the reported studies are given in Appendix II.

II. Experimental Apparatus

II.1. Introduction

In order to develop a clearer understanding of the physics and chemistry ocurring in a plasma processing environment, it is essential that the important discharge parameters be well characterized during the related experimental research. Various macroscopic parameters such as gas mixture ratios, gas flow rates, gas pressure, discharge power, and reactor geometry must be accurately measured and controlled. In addition, the identity and relative density of the various species present in the plasma and at the plasma-surface interface (atoms, molecules, radicals, positive ions, and negative ions) must be determined. This section will describe the experimental systems constructed and, along with two previously existing systems, used in our research efforts under this contract. These systems were designed to provide as much information as feasible about the plasma parameters mentioned above and several aspects of this design philosophy will be discussed in what follows.

Control and measurement capabilities for the macroscopic parameters (flow rate, pressure, etc.) were incorporated into the system design using commercially available components (thermal flow meters, capacitance manometers, etc.). Fortunately, a number of these items became commercially available just as we were designing our first experimental system. These components have been invaluable in making reliable parameter measurements and in controlling the discharge parameters, thus making reproducible experimental results possible. The details of these components will be discussed in the following subsection.

Due to the previous expertise of members of our research group in the application of optical and quadrupole mass spectroscopy to the analysis of gas discharges and due to the appropriateness of the use of these techniques in the experiments we wished to perform, we choose to utilize these analysis methods in our studies. Optical emission spectroscopy can be used to give information about the excited atomic, molecular, and radical species present in the volume of the gas discharge. This information, when correlated with changes in the macroscopic discharge parameters, can yield valuable information about the important plasma reaction mechanisms.

While optical emission spectroscopy is relatively easy to incorporate and use, it does not yield information about the ground state neutral particles or the ionic species present in the discharge. Quadrupole mass spectroscopy has shown itself to be an essential tool for studying these species. However, the incorporation of quadupole mass spectroscopic capabilities and the analysis of the resulting experimental data is not straightforward. Because of the unique constraints imposed by the sampling of neutrals and ions from an active gas discharge, it was necessary for our experimental systems to be designed around this analysis technique, rather than simply utilizing a commercial mass spectrometer attached to a predesigned reactor vessel. Through this approach, we were able to design systems which yield reliable, reproducible information about the various neutral and ionic species present at the plasma-surface interface.

In our plasma volume synthesis studies, these measurements reflect the species concentrations present in the volume of the discharge. While the determination of absolute species densities is not possible in these experiments, the qualitative change in density between various species present as a function of changes in the macroscopic parameters can be investigated. These measurements can again be used to help determine the main volume plasma reactions occuring.

In the case of the plasma processing of material surfaces (polymer film deposition, etching, semiconductor anodization, etc.), it is possible to accurately determine the species bombarding the processed surface by extracting these species through a hole in the surface and then mass analyzing the resulting particle beam. This data yields information about the various species participating in the surface reactions involved in the processing.

In the course of our investigations, two main experimental systems incorporating optical and mass spectroscopy were designed, constructed, and utilized. During the first year of this project, a high vacuum system was constructed for use in the plasma volume synthesis studies. It was decided to construct a system for this purpose first, since the necessary personnel were immediately available for these studies. This system was subsequently used for one and a half years on this aspect of our project. During the third year of the project, this system was modified for use in plasma surface modification studies and is currently being used in this context.

After satisfactory performance of the first system was confirmed, construction was begun on a second, ultra high vacuum system for surface modification studies. Design and construction on this system was completed during the third year of the project and it is currently being used for studies on GaAs anodization and metal film deposition and for measurements related to the modeling of rf discharge sheath properties. This system incorporates several features of the first system, but is an all metal system with ultra high vacuum capabilities. It was decided that this sytem should be designed to allow for maximum experimental flexibility. Hence, the metal reactor chamber has several access ports allowing for the attachment of different combinations of electrical feedthroughs, gas inlets, and viewing windows. It was also determined that the eventual use of beam studies to simulate portions of the plasma environment would be essential for understanding the complicated plasma-surface reactions occuring. With this application in mind, the system incorporated access ports for the attachment of particle (electron, ion, and neutral) and photon beam sources. The pumping requirements of the system were also planned to incorporate these beam studies.

II.2. Plasma Volume Synthesis System

A block diagram of the first system, constructed for the plasma volume synthesis studies, is shown in figure II.2.1. There are five main subsystems in this diagram: (1) the reactor tube (#1 and #2), (2) the mass spectrometer system (#13, #14, and #15), (3) the optical spectroscopy system (#5), (4) the data acquisition system (#16 through #21), and (5) the macroscopic parameter measurement and control system (#3, #4 and #6 through #12). The latter three subsystems will be discussed first, since all our experimental systems incorporated the same general design for these subsystems.

The discharge was powered by an inductively coupled rf excitation source. This source consisted of a 250 watt, 13.56 MHz rf generator (#12). The power from this supply was coupled through a matching network (#10) into a copper induction coil (#9) wound around the one inch diameter glass reactor tube (#1). Power levels were monitored with a wattmeter (#11) and the induction coil voltage was measured with an oscilloscope. The rf power supply was operated in a pulsed mode during some of the optical emission studies. The induction coil could be placed either upstream or downstream of the sampling orifice in order to analyze different regions of the discharge.

The reactor tube was pumped by a chemical resistant 195 l/min rotary vane vacuum pump (#8). A liquid nitrogen cold trap (#7) was incorporated into the pumping system when required to collect

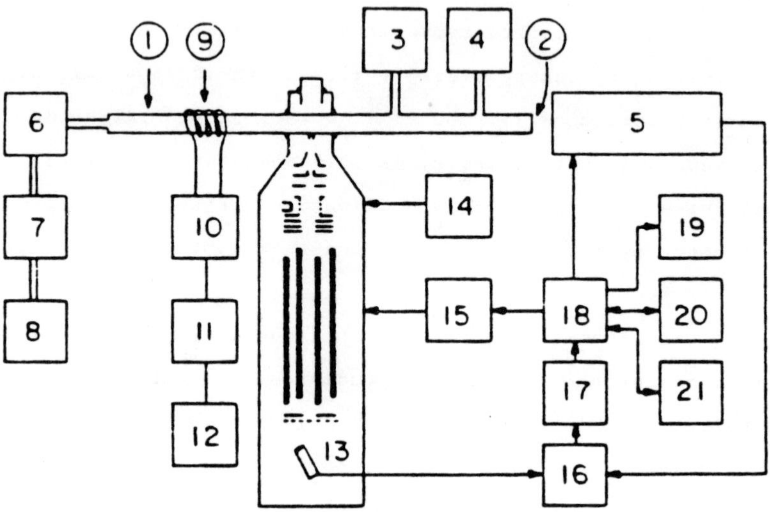

Figure II.2.1: Schematic diagram of plasma volume synthesis system. Refer to the text for a description of the various components.

hazardous or corrosive chemicals. An electropneumatic exhaust valve (#6) and exhaust valve controller was used to control the pumping speed and, hence, the discharge pressure. The pressure was monitored with a temperature controlled 1 torr capacitance manometer (#3). The flow rates and flow ratios (for gas mixtures) were measured and controlled by a three channel flow control system (#4) utilizing thermal flow sensors and electromagnetic valves. This sophisticated pressure and flow control system allowed us to accurately and reproducibly monitor and control the gas pressure and flow rates in the reactor tube. Previous studies in our group and elsewhere had shown that such a control system was essential for reliable experimental measurements.

The optical emission studies were performed by sampling the light emitted from the plasma through a fused silica window (#2) placed on the upstream side of the reactor tube. The emission was spectrum analyzed with a 0.5 m monochromator (#5) and detected with a thermoelectrically cooled photomultipler tube. The spectral range of this detection scheme was 200 nm to 850 nm. This system had sufficient resolution to study the rotational-vibrational structure of the molecular emission bands observed.

Both the emission spectroscopy and mass spectroscopy systems were operated in a pulse counting mode. The pulse counting technique has the advantage of greater dynamic range and increased sensitivity. In addition, because of detector biasing requirements, the pulse counting mode leads to a greatly increased sensitivity and simplified system design when performing negative ion mass spectroscopy. The pulses were amplified (#16), discriminated (#17), and then counted by a microcomputer controlled multichannel scaling data acquisition system (#18). This data acquisition system allowed for subsequent data manipulation, storage, and output to appropriate peripherals (#19, #20, and #21). Computer software for this system was developed to do automatic peak searching and for the plotting of histograms.

The details of the interface region between the reactor tube and the high vacuum mass spectrometer chamber are shown in figure II.2.2. The connection of the reactor tube to the high vacuum system must be tight enough to provide a base pressure of 10^{-8} torr, but also easy to remove. Since plasmas of organic compounds readily generate polymer and carbon layers on the tube wall, frequent disassembly for tube cleaning was required. This connection was accomplished through the use of an interface block with modified Cajon "Ultra-Torr" style viton o-ring seals as indicated. This assembly, with the attached discharge tube, could be removed from the cone enclosing the top of the mass spectrometer chamber for easy cleaning. A viewport was also incorporated into the design to allow for visual access to the sampling orifice.

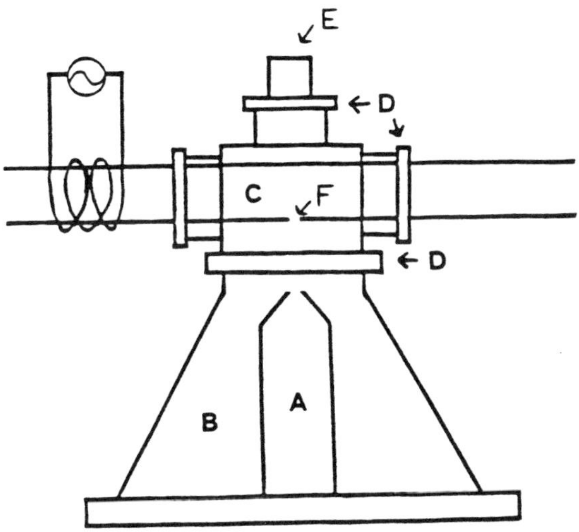

Figure II.2.2: Reactor tube-mass spectrometer chamber interface region.

A - Quadrupole mass spectrometer assembly

B - Stainless steel cone on top of mass spectrometer chamber

C - Aluminum block with modified "Ultra-Torr" style o-ring seals at points D

D - "Ultra-Torr" style seals to top cone (2.5"), reactor tube (1"), and viewing port (1")

E - Viewing port

F - 50 μm to 150 μm sampling orifice

The sampling orifices used were 50 μm to 150 μm in diameter. They were produced by sandblasting from the outside of the discharge tube while a Tesla coil is placed on the inside of the tube directly across from the area being sandblasted. When the glass is thinned to 20 μm to 40 μm, the Tesla coil arcs through the glass, producing a hole. By varying the time of sandblasting after the Tesla coil arcs through from between 10 to 30 seconds, a hole 50 μm to 150 μm in diameter and 15 μm to 30 μm thick can be reliably produced. The resulting hole is round with a cone shaped opening tappering in toward the inside of the discharge tube.

The design of the high vacuum chamber region immediately below the sampling orifice and the placement of the quadrupole mass spectrometer with respect to the orifice are important factors in determining the fate of the ionic and neutral particles between the sampling orifice and the quadrupole mass filter. The system must be designed to maximize the number of particles which arrive at the mass filter without undergoing collisions between the orifice and the mass filter. In order to aid in the design of these system constraints, calculations were performed to determine the effect of various collision processes on the sampled particle beam entering the mass spectrometer. It was found that the loss of ions and neutrals from the beam due to collisions with neutrals in the beam was the predominant collision process and that 10% to 40% of the beam particles underwent collisions between the orifice and the mass filter. However, it was found that greater than 90% of these losses occured within 1 mm of the orifice. Hence, the distance between the sampling orifice and the mass filter is not a critical parameter in this regard. Of additional concern was the ratio of the flux of particles entering the mass filter and coming directly from the primary sampled beam to the flux of particles entering the mass filter after being scattered off the chamber walls. Calculations showed that the shape and dimensions of the entry region to the mass spectrometer chamber has a significant effect on this ratio, with the ratio increasing (as desired) with increasing chamber dimensions. Due to other considerations, such as placement of the rf induction coil, a compromise design was used which yielded a primary to scattered beam ratio of approximately 2.5.

Details of the mass spectrometer chamber and its associated pumping system are shown in figure II.2.3. The containment vessel for the mass spectrometer is constructed of 304 stainless steel and utilizes "conflat" style metal flanges throughout. A 110 l/sec turbomolecular pump backed by a 95 l/min mechanical pump was used in evacuating this chamber. A turbomolecular pump was chosen because of its ability of maintain a "clean" vacuum while pumping many types of chemicals. An ultimate pressure of 3×10^{-8} torr could be attained in this chamber when the ultimate discharge tube pressure was 0.002 torr. For proper operation of the mass

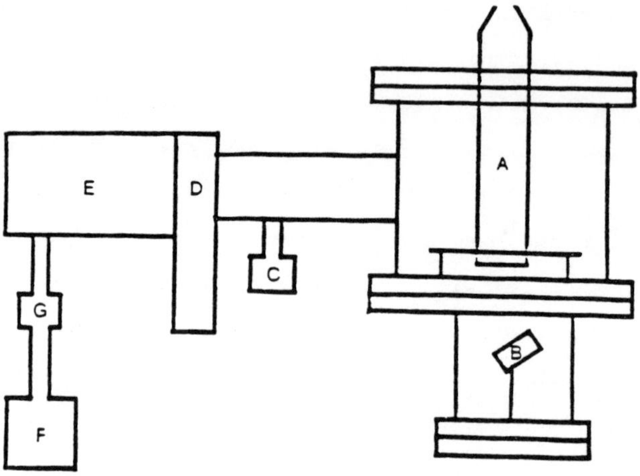

Figure II.2.3: Mass spectrometer chamber and associated pumping system.

A - Quadrupole mass spectrometer assembly

B - Magnetic ion multiplier

C - Ion gauge

D - Gate valve

E - Turbomolecular pump

F - Rotorary vane pump

G - Thermocouple gauge

spectrometer, a vacuum of at least 10^{-5} torr should be maintained inside this chamber during mass analysis. With a sampling orifice of 150 µm and a discharge pressure of 0.1 torr, a pressure of 5×10^{-6} torr could be maintained in the mass spectrometer chamber by this pumping system. Vacuum interlock electronics were constructed in order to shut down the system pumps and quadrupole mass filter electronics in case of a sudden loss of vacuum integrity. The pressure in the mass spectrometer chamber was monitored using an ionization gauge and a thermocouple gauge.

The quadrupole mass spectrometer assembly utilized in this system is shown in figure II.2.4. The mass filter was commercially purchased from Extranuclear Laboratories, Inc. and has 22 cm long, 1.6 cm diameter rods with a floating "ELFS" entrance plate. The quadrupole was originally equipped with an electron beam ionizer and a focusing lens assembly. The ionizer was modifed to incorporate three filaments and to provide a well defined ionization region. A scimmer plate and a split ring plate were added above the ionization chamber. Finally, an electromagnet assembly was mounted on top of the scimmer plate.

This arrangement allows for the mass analysis of either the neutral particles, positive ions, or negative ions extracted through the sampling orifice from the discharge region. When neutral particles produced in the gas discharge tube are to be analyzed, the effluent beam (consisting of neutrals, ions, and

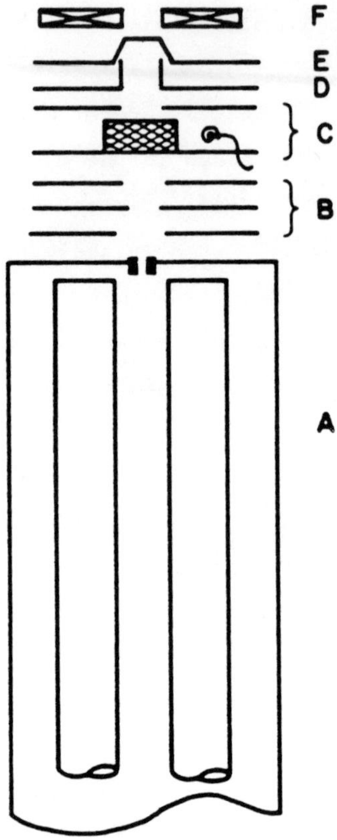

Figure II.2.4: Quadrupole mass spectrometer assembly.

A - Quadrupole mass filter with "ELFS" entrance lens

B - Electrostatic focusing lenses

C - Ionization chamber

D - Split ring

E - Skimmer plate

F - Electromagnet

electrons) enters the mass spectrometer arrangment axially. Electrons are first diverted from the beam using the electromagnet. The remaining neutral and ion beam is then "shaped" by the scimmer plate and the ions are subsequently removed by placing a potential between the two halves of the split ring plate. The remaining neutral beam then passes into the ionization region, after which the ions generated by electron impact ionization of these neutrals are extracted and focused into the quadrupole mass filter.

In the case where the ions produced in the discharge tube are to be mass analyzed, the ionizer is not utilized and, hence, the neutral particles in the beam are not analyzed or detected. The split ring plate is also not used in this case and both halves of this plate are maintained at the same potential. However, the electrons are again removed by the electromagnet. This procedure is especially important when analyzing negative ions, since any electrons which enter the mass analysis system below the scimmer plate can give rise to pseudo-negative ion signals. Finally, the negative or positive ions are selected for mass analysis according to the polarity of the potentials applied to the various elements in the spectrometer assembly. The mass analyzed beam emerges from the quadrupole mass filter and is detected by a magnetic ion multiplier. This ion multiplier has two distinct advantages over the standard "channeltron" design: (1) the multiplier may be disassembled and the dynode strips easily cleaned and rejuvenated and (2) the magnetic field used in the multipler's design helps shield against extraneous electrons which can induce noise signals during negative ion mass analysis.

II.3. Modified System for Surface Studies

During the course of our investigations, the system described in the previous subsection was modified for use in the study of the deposition and removal of material (polymer films, metal films, etc) for surface modification. A schematic diagram of the modified system is shown in figure II.3.1. While the macroscopic parameter measurement and control subsystem was not modified, the reactor tube and reactor tube-mass spectrometer interface region where completely redesigned. Also, additional capabilities were added to the mass spectrometer system itself.

The new discharge tube consists of a four inch diameter glass tube mounted coaxially with the quadrupole mass spectrometer. The reactor tube mounts to the mass spectrometer chamber via a viton o-ring seal, thus facilitating tube removal and cleaning. The top end of the tube is capped by a viton sealed metal flange containing access ports for gas inlet and power feedthroughs. The gas mixture flows into the reactor tube through a port in the top plate and exits to the pump through a one inch glass tube attached

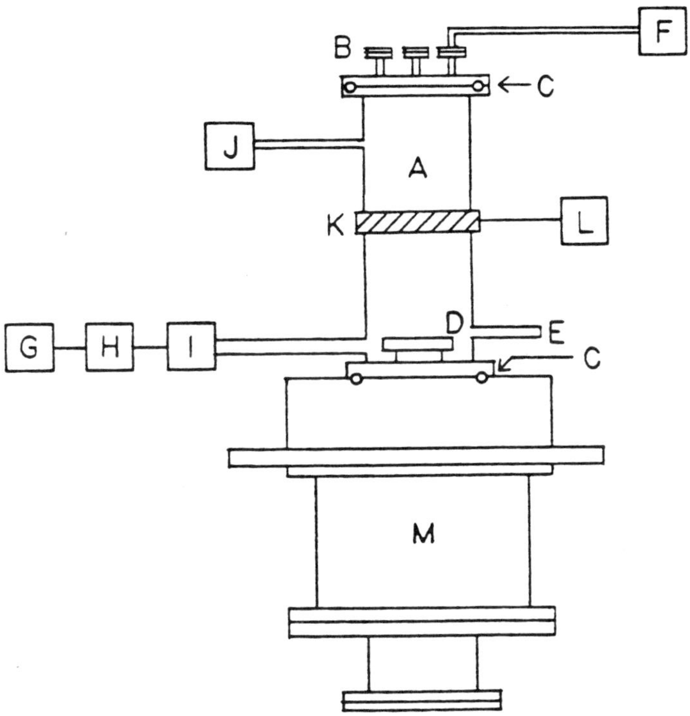

Figure II.3.1: Schematic diagram of modified system for surface studies.

A - Reactor tube

B - Top metal flange

C - Viton o-ring seals

D - Sample holder

E - Quartz window

F - Flow control system

G - Rotary vane pump

H - Liquid nitrogen trap

I - Exhaust valve

J - Capacitance Manometer

K - Capacitive coupling electrode

L - rf power supply and matching network

M - Quadrupole mass spectrometer chamber

to the bottom of the reactor, after flowing over the sample holder. The reactor tube also incorporates a small side tube ending in a fused silica window for use in optical emission spectroscopy. When rf excitation is used, power is typically fed into the discharge region via external capacitive coupling.

Figure II.3.2 shows details of the sample holder arrangement. The sample holder has provisions for the mounting of various substrates such as metal foils and semiconductor wafers. 50 μm to 200 μm holes are placed in the mounted samples through which the neutrals and ions bombarding the surface can be extracted for mass analysis. The sample holder mounts into the opening on the top cover of the mass spectrometer chamber where a high vacuum seal is established between the reactor tube and the mass spectrometer chamber by a viton o-ring seal. This seal also thermally and electrically isolates the sample holder. The sample can be heated by a heater coil attached to the back side of the sample holder. Feedthroughs have been provided in the reactor tube to allow for biasing of the sample holder and for connections to a quartz crystal microbalance which can be attached to the sample holder. The addition of a quartz crystal microbalance allows us to monitor the deposition or removal rates during plasma processing.

The mass analysis system has been modified to incorporate a retarding grid energy analyzer and a motor driven variable

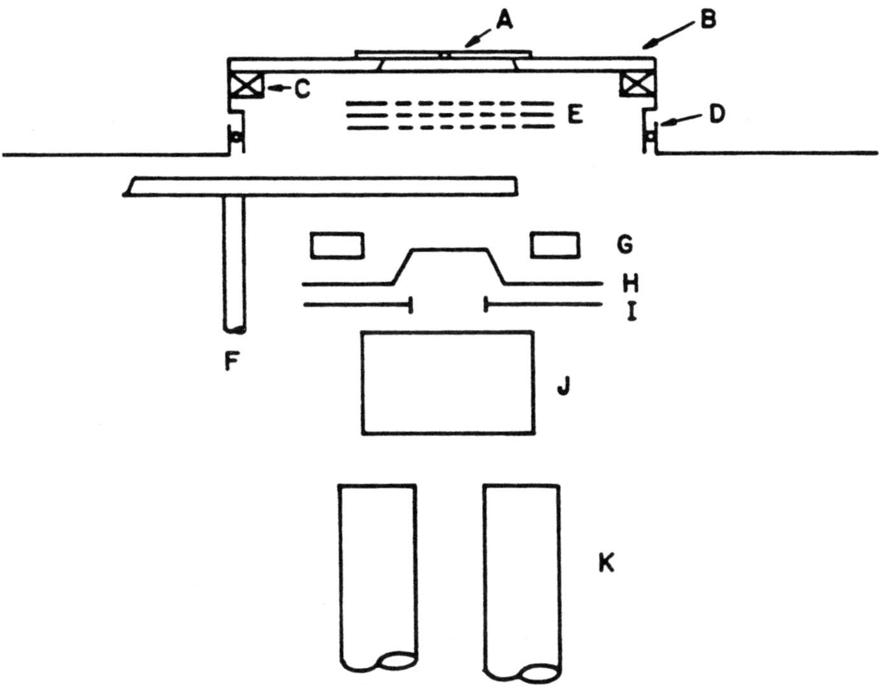

Figure II.3.2: Sample holder and mass spectrometer arrangement.

A - Sample with 50 μm to 150 μm orifice

B - Electrically and thermally isolated sample holder

C - Sample heater

D - O-ring seal for sample holder

E - Ion energy analyzer

F - Beam chopper

G - Electromagnet

H - Skimmer plate

I - Split ring

J - Ionizer-lens assembly

K - Quadrupole mass filter

frequency beam chopper, as shown in figure II.3.2. The energy analyzer provides for a rough determination of the energy distribution of the various ionic species bombarding the sample surface. This information is useful for measuring changes in the bombarding energy as a function of species' mass, discharge pressure, and disharge power. The addition of the beam chopper enhances our sensitivity by a factor of 10. This is very useful in the detection of weak signals arising from short lived radical species. In addition, the chopper can be used to select that portion of the signal arising solely from the primary unscattered ion or neutral beam.

This modified system has a high vacuum base pressure in both the reactor tube and the mass spectrometer chamber of 5×10^{-8} torr. Because of the system design, it has been used for studies involving the deposition and etching of polymer films on various substrates. The design provides for the easy removal of the processed samples and for the frequent cleaning of the reactor tube which is required in this "dirty" environment. The versatility of the metal flange on the top of the reactor tube allows us to reconfigure the system in order to use either dc excitation or internally capacitive coupled rf excitation in place of externally coupled rf excitation.

II.4. Ultra High Vacuum Surface Modification System

Figure II.4.1 contains a schematic diagram of the ultra high vacuum system designed for plasma-surface and beam-surface interaction studies. This system is similar in design to the modified system described in the previous subsection, except that the reactor chamber is constructed of 304 stainless steel and utilizes all metal "conflat" flanges. This reactor has 16 access ports of various sizes located around the cylindrical walls of the chamber. These ports provide for access to the reactor chamber for sample interchange and for attachment of (1) pressure sensors (a capacitance manometer and an ionization gauge), (2) a gas inlet line (from a flow control system as described previously), (3) a gas outlet line (connected to a 195 l/min mechanical rotary vane pump), (4) glass and quartz windows (for viewing and for optical spectroscopy), (5) electrical feedthroughs (for rf and dc excitation power, sample holder biasing, and quartz crystal microbalance connections), and (6) an interconnect line between the reaction chamber and the mass spectrometer chamber (so that both chambers can be pumped simultaneously by the turbomolecular pump attached to the mass spectrometer chamber and backfilled simultaneously with dry nitrogen). A dry nitrogen purge capability is used to maintain system cleanliness during system maintenance and sample interchange. In addition, a ten inch access port is used for the attachment of a gate valve and an 8

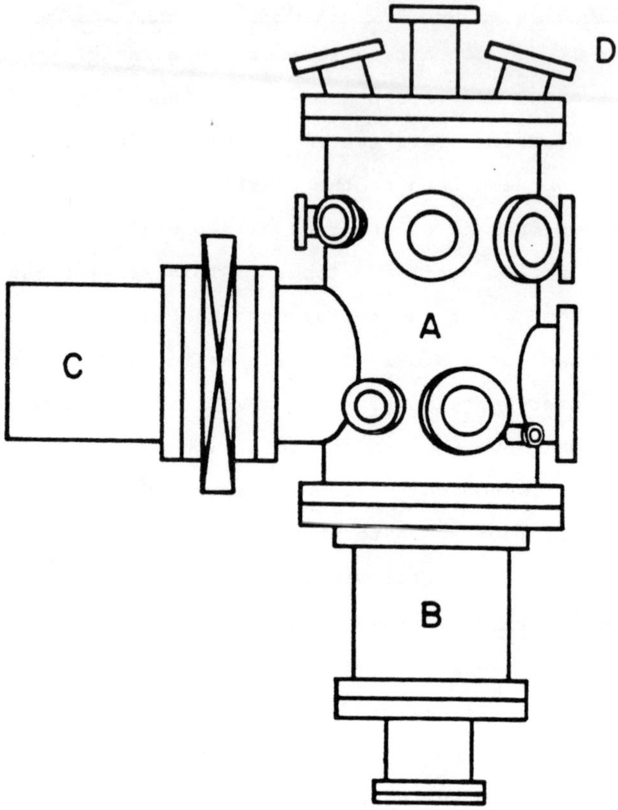

Figure II.4.1: Schematic diagram of ultra high vacuum surface modification system.

A - Main reactor chamber

B - Quadrupole mass spectrometer chamber

C - Cryopump

D - Top flange for mounting beam sources

inch, 4000 l/sec cryopump. This pump provides the pumping speed required to obtain ultra high vacuum conditions of 10^{-10} torr in the reactor chamber and for maintaining a low background pressure in the reactor chamber during the beam studies.

The top flange on the reactor chamber contains access ports for the future attachment of various beam sources and, if desired, a sample manipulator. These ports are aligned so as to converge on the surface of the sample mounted on the sample holder. The sample holder assembly and the quadrupole mass spectrometer system are of the same design as described in the previous subsection.

III.Preparation and Characterization of Plasma Polymerized Polyvinylferrocene

This section describes work carried out primarily by Professor Evans and his group in which the goals were to demonstrate that polymeric films of predictable structure could be prepared by plasma processing. The vinylferrocene monomer was chosen for this study because it afforded a property which other plasma polymerized materials do not: electrochemical activity. As such, films prepared by plasma deposition from neat vinylferrocene vapor could be examined not only by the usual spectroscopic techniques, but also equilibrium and dynamic electrochemical techniques. Furthermore, the linear polymer was available from commercial sources, so that in all experiments, comparisons could be readily drawn between the linear material and the plasma synthesized polymers.

This section of the report is organized into several subsections each of which deals with a different aspect of the overall project. The first two subsections deal with the application of coalmetric titration procedures (III.1) and ellipsometric thin film thickness determinations (III.2) as they apply to the study of plasma polymerized vinylferrocene (PPVF). Although these are both fairly standard procedures, there are some subtle aspects as applied here. Next, an experimental subsection

(III.3) details the preparation and characterization procedures used. The results of both surface and bulk analytical spectroscopies are elaborated in the subsequent subsection, after which there is a subsection dealing with the thin film analysis of PPVF by X-ray photoelectron spectroscopy and secondary ion mass spectrometry (III.5). The final two subsections are devoted to a discussion of the thermodynamics (III.6) and kinetics (III.7) of electron transfer within PPVF compared to its conventional, linear analogue: linear polyvinylferrocene (LPVF).

III.1. Coulometry for Characterization of Electroactive Polymer Modified Electrodes

Cyclic voltammetry is the most extensively used technique for electrochemical characterization of electroactive polymer modified electrodes. It is an excellent technique for qualitatively examining a wide range of properties. However, its utility in quantitating electrochemical coverage and evaluating thermodynamic properties (concentration - potential relationships) of polymer modified electrodes is hindered due to the tedious nature of extracting these data from generated cyclic voltammograms.

To date, characterization of electroactive polymer modified electrodes using coulometric methods of analysis has been virtually ignored. This is rather surprising, since coulometric methods can offer certain advantages over other electrochemical methods, and coulometric methods are simple to implement, both in terms of hardware and execution. Some of the advantages of coulometric techniques will be discussed, along with comparisions to other electrochemical techniques, such as cyclic voltammetry.

Determination of Electrochemical Coverage

Consider an electroactive modifier immobilized on an electrode surface which undergoes the chemically reversible electron transfer reaction shown below:

$$R \rightleftharpoons O + ne^- \qquad E^o = E^o_{O/R} \qquad (1)$$

where $E^o_{O/R}$ is the standard potential for the redox process. It is assumed that the reaction obeys the Nernst equation (2),

$$E = E^{o}_{O/R} + \frac{RT}{nF} \ln \left(\frac{a_O}{a_R} \right) \qquad (2)$$

where a_O and a_R are the activities of the oxidized and reduced forms of the electroactive modifier, respectively. Equation (2) may be rewritten to express the concentration of the electroactive modifier

$$E = E^{o}_{O/R} + \frac{RT}{nF} \ln \left(\frac{\gamma_O C_O}{\gamma_R C_R} \right) \qquad (3)$$

where γ_O and γ_R are the activity coefficients for the oxidized and reduced forms, and C_O and C_R are the surface concentrations of the redox forms of the electroactive modifier. Expanding equation (3) gives:

$$E = E^{o} + \frac{RT}{nF} \ln \left(\frac{\gamma_O}{\gamma_R} \right) + \frac{RT}{nF} \left(\frac{C_O}{C_R} \right) \qquad (4)$$

Defining $E^{o'}_{O/R}$ ad the formal potential by:

$$E^{o'}_{O/R} = E^{o}_{O/R} + \frac{RT}{nF} \ln \left(\frac{\gamma_O}{\gamma_R} \right) \qquad (5)$$

Substitution into (4) gives:

$$E = E^{o'}_{O/R} + \frac{RT}{nF} \ln \left(\frac{C_O}{C_R} \right) \qquad (6)$$

At 25° C:

$$E = E^{o'}_{O/R} + \frac{0.059}{n} \log \left(\frac{C_O}{C_R} \right) \qquad (7)$$

Γ_T is defined as the total surface coverage of the electroactive modifier (in moles-cm^{-2}) and Γ_O and Γ_R are defined as the surface coverage at any potential of the oxidized and reduced form of the electroactive modifier respectively. By the requirement of conservation of total quantity of electroactive modifier

$$\Gamma_T = \Gamma_O + \Gamma_R \tag{8}$$

equation (7) may be rewritten to express the potential as a function of Γ_O and Γ_R by dividing these quantities by the thickness (d) of the modifier layer:

$$C_O = \frac{\Gamma_O}{d}, \quad C_R = \frac{\Gamma_R}{d} \tag{9}$$

This gives

$$E = E^{o'}_{O/R} + \frac{0.059}{n} \log \frac{\Gamma_O/d}{\Gamma_R/d} \tag{10}$$

which reduces to

$$E = E^{o'}_{O/R} + \frac{0.059}{n} \log \frac{\Gamma_O}{\Gamma_R} \tag{11}$$

The above equation (11) is similar to those derived by others to describe potential - concentration relationships of immobilized electroactive species [1]. It is important to note that equation (11) does not include a term to describe possible activity effects due to interactions between electroactive centers in the modified layer. For the discussion that follows, it will be assumed that the centers do not intereact and equation (11) properly describes the Nernst equation for the reaction in equation (1).

The quantitation of electrochemical coverage of the electroactive modifier might be achieved by a variety of potential perturbation techniques. Care must be taken in choosing the potential window for the potential perturbation, as shown below:

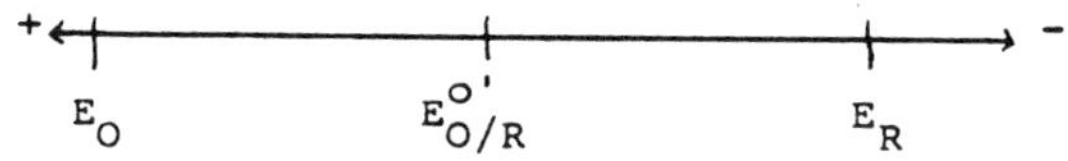

$E^{o'}_{O/R}$ is the formal potential of the electroactive couple, E_O must be sufficiently anodic to ensure that the electroactive modifier is in the completely oxidized form and E_R is sufficiently cathodic to ensure that the electroactive modifier is in the completely reduced state. By equation (11), $E_O = E^{o'}_{O/R} + \frac{180mV}{n}$ and $E_R = E^{o'}_{O/R} - \frac{180mV}{n}$ to convert the modifier layer to 99.9% in the oxidized and reduced form respectively. To quantitate the surface coverage, the potential is changed from E_R to E_O (or *vice versa*) while measuring the charge passed due to the potnetial perturbation. It is further assumed that the electroactive modifier is indefinitely stable and that no other faradaic processes occur except for the redox of the electroactive modifier. In this case (assuming equilibrium) the total charge passed (Q_T) due to switching the potential (either by potential step or potential sweep) from $E_R \rightarrow E_O$ (or *vice verse*) will equal the sum of the double layer charging (Q_{dl}) and the charge due to electrochemical conversion of $R \rightarrow O$ ($Q_{T,O/R}$)

$$Q_T = Q_{dl} + Q_{T,O/R} \qquad (12)$$

where $Q_{T,O/R} = \Gamma_T nFA$, A is the electrode area.

It is also assumed that the double layer charging acts as an ideal capacitor (C_{dl} is independent of potential) so that the double layer charging may be described by equation (13)

$$Q_{dl} = C_{dl}(E_O - E_R)A \qquad (13)$$

where C_{dl} is the electrode capacitance. Typical values for C_{dl} are 15 to 20 μF-cm^{-2} [2]. If $E_O - E_R$ is .36 volts and A is 1 cm, then Q_{dl} is 7.2 μC for $C_{dl} = 20\ \mu f/cm^2$. Assuming a monolayer of a typical organic or organometallic electroactive modifier corresponds to a surface coverage of 5 x 10^{-10} moles-cm^{-2}, $Q_{T,O/R}$ = 48 μC for a one electron transfer reaction. Hence, for monolayer or less coverages, large errors in determining Γ_T would be encountered by a coulometric determination if Q_{dl} is not accounted for. In the above example Γ_T would be overestimated by 15%.

One method to crudely correct Q_T for Q_{dl} would be to measure Q_{dl} in the potential window of interest at the unmodified electrode and subtract this quantity from Q_T to get $Q_{T,O/R}$ directly. A necessary assumption here is that C_{dl} is not altered as a consequence of the modification. This might not be a valid assumption, and C_{dl} may increase or decrease due to the presence of the electroactive modifier [3].

Another approach employing coulometry which allows for correction of Q_{dl} was demonstrated by Stankovich <u>et al</u>. [4]. In this approach, the potential limits of integration (E_O and E_R) are varied and since by equation (12) and (13)

$$Q_T = AC_{dl}(E_O - E_R) + nFA\Gamma_T \qquad (14)$$

a plot of Q_T <u>vs</u>. $(E_O - E_R)$ can be employed to determine C_d and Γ_T.

It is stressed that E_O and E_R must always be large enough to ensure complete conversion from one redox form to the other.

Cyclic voltammetry is very useful for quantitating surface coverage when the double layer charging contributes significantly to the current or charge passed with a linear potential sweep perturbation. By extending the potential perturbation limits beyond that required for total conversion, examination of the charging current in a region removed from the faradaic conversion of the modifier is possible. The double layer charging current may be extrapolated through the region of observed faradaic current and the charging current contribution to the total current may thus be subtracted out. Quantitation of surface coverage is achieved by integrating the area defined by the current/potential curve and the extrapolated base line charging current by graphical integration methods. This approach has found wide acceptance for quantitating surface coverage of monolayer or lower coverages [5-8].

Even though large errors are encountered employing coulometric methods for surface quantitation without correction for double layer charging, the relative error decreases as coverage increases, and may become insignificant for multilayer coverages. Assuming $E_O - E_R = 0.36V$ and $C_{dl} = 20\ \mu f/m^2$, the calculated error is only 1.5% for $\Gamma_T = 5 \times 10^{-9}$ moles-cm^{-2} and 0.1% for $\Gamma_T = 5 \times 10^{-8}$ moles-cm^{-2} (these represent approximately 10 and 100 equivalent monolayers, respectively). So even for relatively thin electroactive polymer modified electrodes, a high degree of accuracy is possible using the coulometric technique. The accuracy expected from quantitation by cyclic voltammetry is no better due to graphical methods of integration employed. These integration methods are also tedious and time consuming compared to coulometric methods.

It was assumed in the above discussion that the electroactive modifier is at all times in equilibrium with the electrode. That is, the conversion kinetics are fast with respect to the experimental time frame. An important advantage that can be realized with linear sweep coulometry is verification of what time frame is needed to establish equilibrium, since at equilibrium the volt-coulogram will be independent of sweep rate. In the absence of any background faradaic process, a simple test to establish equilibrium would be to reduce the sweep rate until the volt-coulograms are superimposable over each other. This simple test is not possible with linear sweep voltammetry since the current is directly proportional to the sweep rate. One would have to examine the cyclic voltammetry as a function of sweep rate, and graphically integrate each wave at each sweep rate to verify that the integrated charge is independent of sweep rate. This method would be excessively time consuming. An alternative approach would be to electronically normalize the cyclic-voltammograms by attenuating the current by the inverse of the sweep rate. When the normalized cyclic-voltammograms are found to be independent of sweep rate, equilibrium can be assumed, and only one integration is needed for quantitation.

In the event that the conversion kinetics are so slow as to make linear sweep techniques inconvenient, one may simply step the potential from $E_R \rightarrow E_O$ (or vice vera) and hold the potential at E_O until the charge accumulation (or current) decays to zero. This method is more easily implemented by coulometry. This potential step method is also very useful for quantitation of surface coverage even in the presence of a chemically irreversible solution background reactions. Consider an irreversible reaction of dissolved A and B:

$$A \longrightarrow B + e^- \qquad E = E_{A/B} \qquad (15)$$

where $E_{A/B}$ is defined as the potential where the onset of oxidation of A to B occurs. For sake of argument, let $E_{A/B}$ lie somewhere between $E^{o'}_{O/R}$ and E_O. If the potential is stepped from $E_R \rightarrow E_O$, and held at E_O for some time interval t, t being sufficiently long for total conversion, the accumulated charge is given by (16):

$$Q_{T(a)} = Q_{T,O/R(a)} + Q_{A/B(a)} \qquad (16)$$

where $Q_{T(a)}$ is the total accumulated anodic charge, $Q_{T,O/R(a)}$ is the anodic charge due to conversion of $R \rightarrow O$, and $Q_{A/B}$ is the anodic charge due to oxidation of $A \rightarrow B$. It is assumed that the double layer charging is small relative to the charge due to conversion of $R \rightarrow O$ and hence is not considered. If the potential is then stepped back to E_R and held at E_R until charge accumulation ceases, the total charge will be given by:

$$Q_T = Q_{T(a)} - Q_{T,O/R(c)} = Q_{A/B(a)} \quad (17)$$

where $Q_{T,O/R(c)}$ is the cathodic charge due to conversion of O T R. Subtracting (17) from (16) gives $Q_{T,O/R(a)}$ directly, which is the charge due to total conversion of R → O. By varying the anodic hold time, verification of equilibrium is achieved by the constancy of the calculated $Q_{T,O/R(a)}$. Hence, even in the case of irreversible background solution faradaic processes, coulometry can be used to accurately measure the surface coverage of electroactive polymer modified electrodes.

Coulometry for Evaluation of Thermodynamics of Electroactive Polymer Films

Recently there has been considerable interest in studying the electrochemical thermodynamics of surface bound electroactive modifiers. Most of the work published in the literature has dealt with monolayer coverages or less of electroactive modifiers [6,9]. It has been found, in some instances, that the concentration - potential relationships do not follow the Nernst equation as written in equation (11). Several have invoked an interaction parameter to account for the observed non-idealities [10-13]. This interaction parameter is an activity coefficient(s) which varies exponentially with the surface coverage of the modifier and the fraction of the modifier in the reduced and oxidized form. Strictly speaking, theories that have been derived which include this interaction parameter are applicable only to monolayer of less coverages of modifier [14] and cannot be assumed to be applicable to multilayer coverages, even though attempts to apply the theory to electroactive polymer modified electrodes have been made [15]. Even in the case of monolayer coverages or less, rigorous tests of the validity of the theory have not been demonstrated. These theories predict that the interaction between electroactive centers should become less important as the surface coverage decreases while the value of the interaction parameter remains constant. The author knows of only one study which examined this by varying surface coverage [6]. In that study the interaction parameter was not constant as a function of coverage.

In general, transient electrochemical techniques (e.g. cyclic voltammetry and chronopotentiometry) have been used to examine the thermodynamics of electroactive modified electrodes. In certain cases, the interaction parameter had different values, depending upon whether the modifier was being oxidized or reduced [15]. This seems to violate thermodynamic theory and suggests that the system was not in equilibrium during the measurement.

This author felt that coulometric methods were ideally suited for evaluation of thermodynamic properties for several reasons. Firstly, the relationship that one is trying to establish is the ratio of the amount of the redox couple in its oxidized and reduced form as a function of potential. Coulometry as a function of potential achieves this directly as shown below:

$$\Gamma_O(E) = nFAQ_a(E) \qquad (18)$$

where $\Gamma_O(E)$ is the amount of the oxidized form at any potential for a perturbation from $E_R \rightarrow E_O$, (E) is a potential between E_O and E_R and

$$\Gamma_R(E) = \Gamma_T - \Gamma_O(E) \tag{19}$$

Hence, by using either linear potential sweep coulometry or the double potential step coulometry discussed previously, one may find Γ_O and Γ_R at any potential. Secondly, it can be easily verified that the measurement is being performed at equilibrium as discussed in the last section. Also, if one uses cyclic coulometry, equilibrium may be verified by the superimposability of the forward and reverse sweep.

The Nernst equation may be written as a function of charge at any potential and the total charge. Since it was the purpose of these coulometric experiments to determine the thermodynamics of PPVF film conversion, an activity term to describe possible interactions between ferrocene sites is included in the Nernst equation

$$E = E^{o'} + \frac{RT}{nF} \ln \gamma'_{O/R} \frac{\Gamma_O}{\Gamma_R} \tag{20}$$

Here $\gamma'_{O/R}$ is a generalized activity term where

$$\gamma_{O/R} = f(\Gamma_O, \Gamma_R) \tag{21}$$

Substituting (18) and (19) into (20)

$$E = E^{o'} + \frac{RT}{nF} \ln \left\{ \gamma'_{O/R} \left(\frac{Q_{a(E)}}{Q_{T,O/R} - Q_{a(E)}} \right) \right\} \tag{22}$$

and at 25° C

$$E = E^{o'} + \frac{0.059}{n}\left\{\log(\gamma'_{O/R}) + \log\left(\frac{Q_{a(E)}}{Q_{T,O/R} - Q_{a(E)}}\right)\right\} \quad (23)$$

If there are no interactions between electroactive centers, $\gamma'_{O/R}$ is equal to one. If there are interactions between electroactive centers, $\log\left(\frac{Q_{a(E)}}{Q_{T,O/R} - Q_{a(E)}}\right)$ vs. E may be used to find the functional form of the dependence of $\gamma'_{O/R}$ on Γ_O and Γ_R.

References

1. Bard, A. J. and Faulkner, L. R., Electrochemical Methods. Fundamentals and Applications, John Wiley & Sons, New York, 1980, p. 522.

2. Sawyer, D. T. and Roberts, Jr., J. L., Experimental Electrochemistry for Chemists, John Wiley & Sons, New York, 1974, p. 70.

3. Ref. 1, p. 535.

4. Stankovich, M. T. and Bard, A. J., J. Electroanal. Chem., 86, 189 (1978).

5. Rocklin, R. D. and Murray, R. W., J. Phys. Chem., 85, 2104 (1981).

6. Lenhard, J. R. and Murray, R. W., J. Amer. Chem. Soc., 100, 7870 (1978).

7. Sharp, M., Petersson, M. and Edstrom, K., J. Electroanal. Chem., 95, 123 (1979).

8. Laviron, E., J. Electroanal. Chem., 101, 19 (1979).

9. Brown, A. and Anson, F. C., J. Electroanal. Chem., 92, 133 (1978).

10. Laviron, E., J. Electroanal. Chem., 52, 395 (1974).

11. Brown, A. P. and Anson, F. C., Anal. Chem., 49, 1589 (1977).

12. Conway, B. E., Angerstein-Kozlowska, H. and Ho, F. C., J. Vac. Sci. Technol., 14, 351 (1977).

13. Angerstein-Kozlowska, H., Klinger, J. and Conway, B. E., J. Electroanal. Chem., 75, 45 (1977).

14. Ref. 1, p. 517.

15. Daum, P. and Murray, R. W., J. Phys. Chem., 85, 389 (1981).

III.2 Ellipsometric Measurement of PPVF Film Thickness and Refractive Index

Ellipsometry is a technique that allows for the determination of the optical properties of a surface or film between two media and is based on exploiting the change in polarization that occurs as a beam of polarized light is reflected from or transmitted through the surface or film. Two factors make ellipsometry particularly attractive for thin film characterization: (1) its essentially non-perturbing character (with proper choice of intensity and wavelength of the incident light beam) and (2) its ability to accurately measure the optical properties of thin films. Depending on the sensitivity of the ellipsometer used, the refractive index and thickness of films from a few angstroms to several thousand angstroms may be accurately measured. It is because of this ability that ellipsometry has become a valuable tool in the study of surfaces and thin films. Some of the applications of ellipsometry which have been reported include: measurement of properties of dielectric films on silicon [1,2], study of adsorption of polymers from solution [3,4], gas adsorption studies [5], studies of double layer structure at electrode interfaces [6], *in situ* measurement of polymer film growth rate formed by electrochemical polymerization of phenolic compounds [7], and measurement of polymer film growth rate in the plasma polymerization of styrene [8].

Due to the accuracy of film thickness measurement afforded by ellipsometry, this technique was the method chosen to measure the thickness and optical constants of PPVF films.

Δ and ψ are the measured parameters which are related to the polarizer and analyzer settings (P and A, respectively, for a fixed compensator of $+45^{\circ}$ azimuthal angle only) by the following relationships:

$$\Delta = -2P \quad \pi/2 \tag{1}$$

$$\psi = A \tag{2}$$

In PCSA ellipsometry with the compensator fixed at $+45^{\circ}$, there are two zones or two non-redundant null pairs (of P and A settings) which will satisfy the condition for a null. In the PCSA ellipsometry nomenclature for a compensator set at $+45^{\circ}$, the zones are referred to as zones 2 and 4 and the null pairs as (P_2, A_2) and (P_4, A_4).

The polarizer and analyzer readings in the two zones are related by:

$$P_4 = P_2 + 1/2\pi \tag{3}$$

$$A_2 + A_4 = \pi \tag{4}$$

Even though the values of one null pair may be used to give Δ and ψ, zone averaging is generally used since it reduces and minimizes much of the errors caused by imperfect optical components [9]. In this case, Δ and ψ are given by:

$$\psi = -1/2(A_4 - A_2) \tag{5}$$

$$\Delta = -(P_2 + P_4) \tag{6}$$

The measured values of Δ and ψ depend on the wavelength of light (λ), the angle of incidence (ϕ), the refractive index of the medium ($\hat{n}_3$) and the optical constants of the sample. The medium refers to the environment in which the sample is examined. The optical constants of the sample refer to the refractive index of the substrate ($\hat{n}_3$) in the case of a bare substrate, and to the refractive index of the film ($\hat{n}_2$), thickness of the film (d) as well as $\hat{n}_3$ in the case of a film covered surface. Here, the refractive indices are complex quantities where

$$\hat{n} = n\,(1-iK) \qquad (7)$$

n is the real part of the refractive index and K, the imaginary part, is a measure of absorption by the substrate or film. With the measurement of Δ and ψ, the optical constants of the sample may be calculated.

The equations (and their derivations) which are used to calculate the optical constants of a sample from Δ and ψ may be found in several sources [1, 9-12]. The calculations are rather lengthy, in fact so lengthy that prior to thte advent of the computer, only approximations to the ellipsometric equations were used to calculate optical constants of surfaces and films. Implementation of the calculations by computer has allowed for the practical utilization of the exact ellipsometric equations [13].

For a film free surface, there are two unknown quantities which must be calculated from the measured values of Δ and ψ. These are n_3 and k_3. The other parameters of the optical system are known (e.g. λ, ϕ and n_1). The medium in which the measurements are taken in generally aír, with a refractive index of 1.00. With the two measured quantities, Δ and ψ, n_3 and k_3 may be solved directly.

The situation is more complex with a film covered surface. The complex index of refraction of the film free surface must be known in order to calculate the optical constants and thickness of the film. The optical constant of the film free surface can be determined ellipsometrically as stated above. If the film is non-absorbing ($k_2=0$) then there are only two unknown quantities, d and n_2. In this case a single measurement of Δ and ψ from the film covered surface permits calculation of d and n_2. However, if the film is absorbing ($k_2 \neq 0$) then there are three unknowns (d, n_2, and k_3) and a unique solution cannot be found from one (Δ,ψ) data pair. At least one of the parameters of the optical system must be varied in order to generate a third equation to allow for solving of the three unknowns. There are several approaches that may be used to generate the "third" equation. These include: (1) measurement of Δ and ψ at two or more discrete wavelengths, (2) measurement of Δ and ψ at two or more angles of incidence, (3) changing the refractive index of the medium, (4) analysis of the film deposited on two or more different substrates with different indices of refraction and (5) examination of various thicknesses of the film. The relative merits of these approaches have been discussed by McCrackin and Colson [14].

There are two important considerations when interpreting ellipsometric data to calculate the index of refraction and thickness of non-absorbing films. The first is that the

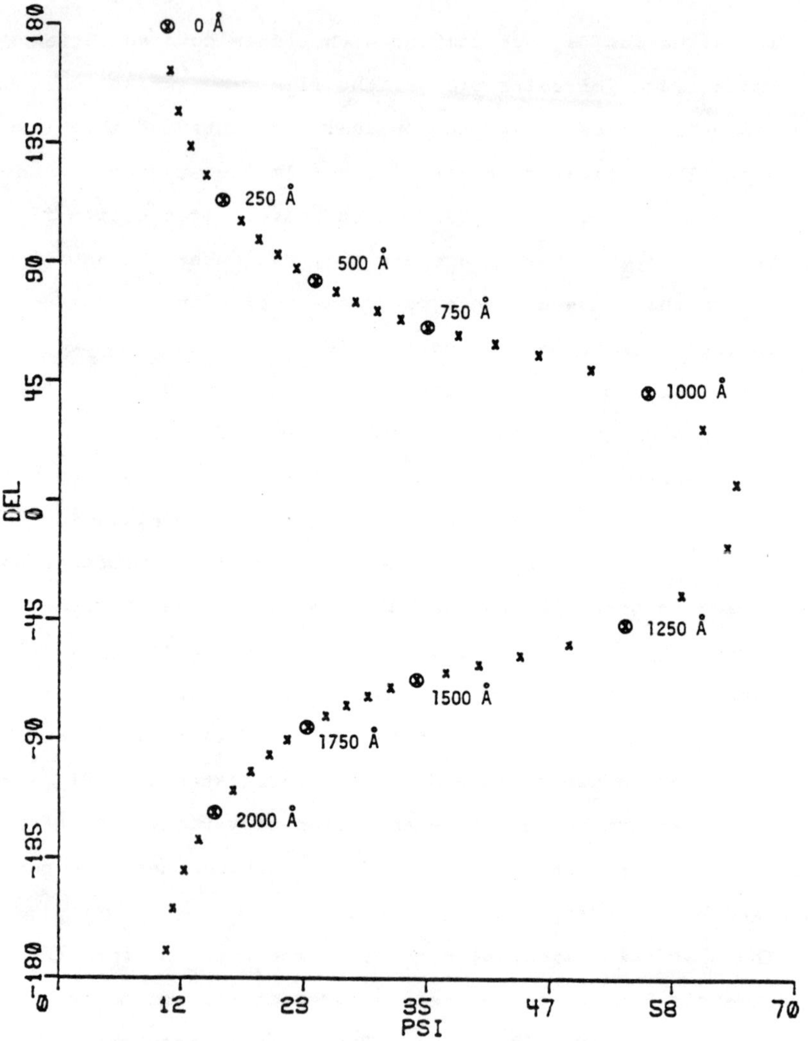

Figure III.2.1: Ellipsometric response function for a film on silicon with a film refractive index of 1.70. Other parameters used are λ = 6328, ϕ = 70.0, and n_1 = 1.00. Each point (x) represents a 50 Å increment.

ellipsometric response function (expressed as (Δ, ψ) data pairs) for a given refractive index of the film is a periodic function of thickness. The periodicity is given by the following relationship

$$\tau = \frac{\lambda}{2n_2}\left[1 - \left(\frac{n_1}{n_2} \sin \phi\right)^2\right]^{-1/2} \qquad (8)$$

where τ is the thickness period and the other variables have been previously defined. The calculated ellipsometric response function for a non-absorbing film with a refractive index of 1.70 on silicon is shown in Figure 1. The values of the other parameters used are λ = 6328 Å, n_1 = 1.00, ϕ = 70.0°, $\hat{n}_3$ = 3.85 (1-0.005 i). τ calculated for the parameters given is 2233 Å. The thickness period is independent of the substrate refractive index. The ellipsometric response function traces over itself for film thicknesses greater than the first cycle length. Hence, the (Δ, ψ) data pair is not unique in the sense that it only can give the thickness as d + mτ where m is a positive integer. This requires knowing the thickness of the film to ≈2000 Å, which usually does not represent a problem. Independent film thickness measurements or film gorwth rate data are generally sufficient to determine m.

Inspection of the shape of the ellipsometric response function shown in Figure III.2.1 reveals that the greatest change in Δ and ψ occurs near the half-period thickness for this particular film on silicon. The greatest accuracy in film thickness measurement is achieved at this point. The maximum sensitivity does not occur at the same point for all film-substrate combinations and is a function of both the film and substrate refractive indices. This can be advantageous as will be shown later. What should be noted, however, is that the accuracy of film thickness measurement depends strongly on the thickness of the film.

The other consideration which must be taken into account when interpreting ellipsometric data is the ability to accurately calculate both the film thickness and the refractive index. For illustrative purposes, Figure III.2.2 shows the calculated ellipsometric response function for films on silicon with refractive indices of 1.65, 1.70 and 1.75. It is seen that the greatest difference in Δ and ψ values between the isorefractive curves occurs near half the period thickness. The isorefractive curves merge for very thin films and for films of thickness close to the period thickness. In many cases random errors, systematic errors, or the resolution of the ellipsometer used do not allow measurement of Δ and ψ to better than a few tenths of a degree accuracy. Hence, isorefractive lines separated by less than a few tenths of a degree may not be resolved and relatively large errors in film thickness calculations are possible. One procedure to reduce the error in film thickness measurement involves calculating the film refractive index of the film for a thickness of half the period thickness. This calculated film refractive index is then fixed in the calculation for measurement of thinner

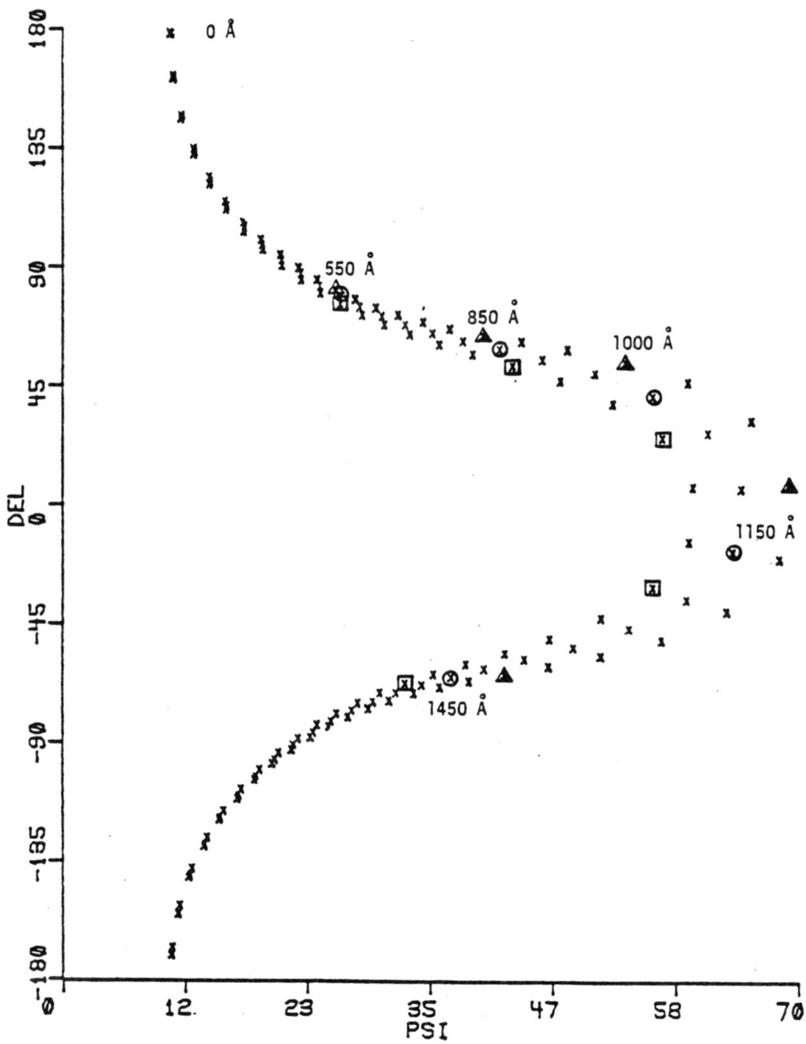

Figure III.2.2: Ellipsometric response function for films on silicon with film refractive indices of 1.65 (△), 1.70 (○) and 1.75 (□). Each point (x) on an isorefractive curve is incremented by 50 Å. Selected film thicknesses are shown adjacent to point marked (△, ○, □) on isorefractive curves. Other parameters are the same as those of Figure 1.

films or films near the period thickness. This procedure can be used only in cases where preparative parameters can be accurately and reproducibly controlled to yield a material having a constant refractive index. If this condition is met, the procedure described greatly reduces the error in film thickness measurement.

Another approach is to examine the films prepared on substrates with different refractive indices, assuming of course that this may be achieved for the system under study. Figure III.2.3 shows the ellipsometric response function for films of refractive indices of 1.65, 1.70 and 1.75 on glassy carbon (refractive index of glassy carbon is 1.864 (1-0.395 i) vide post). It is seen that the maximum sensitivity occurs for films at about one third the period thickness. Hence judicious choice of substrates may aid in determining the thickness and refractive index of films if the films can be formed at the same rate on different substrates during growth. Furthermore, the film thickness may be measured over a wide range of thicknesses without having to assume the refractive index of the film.

Other conditions must be met to properly use an ellipsometric measurement. The substrate must be smooth relative to the wavelength of the incident beam so that the light is not scattered, which would cause partial dipolarization of the light. The effects of surface roughness on ellipsometric measurements have been examined by several investigators [9,15,16]. The substrate must also be flat. If there is a tilt in the sample,

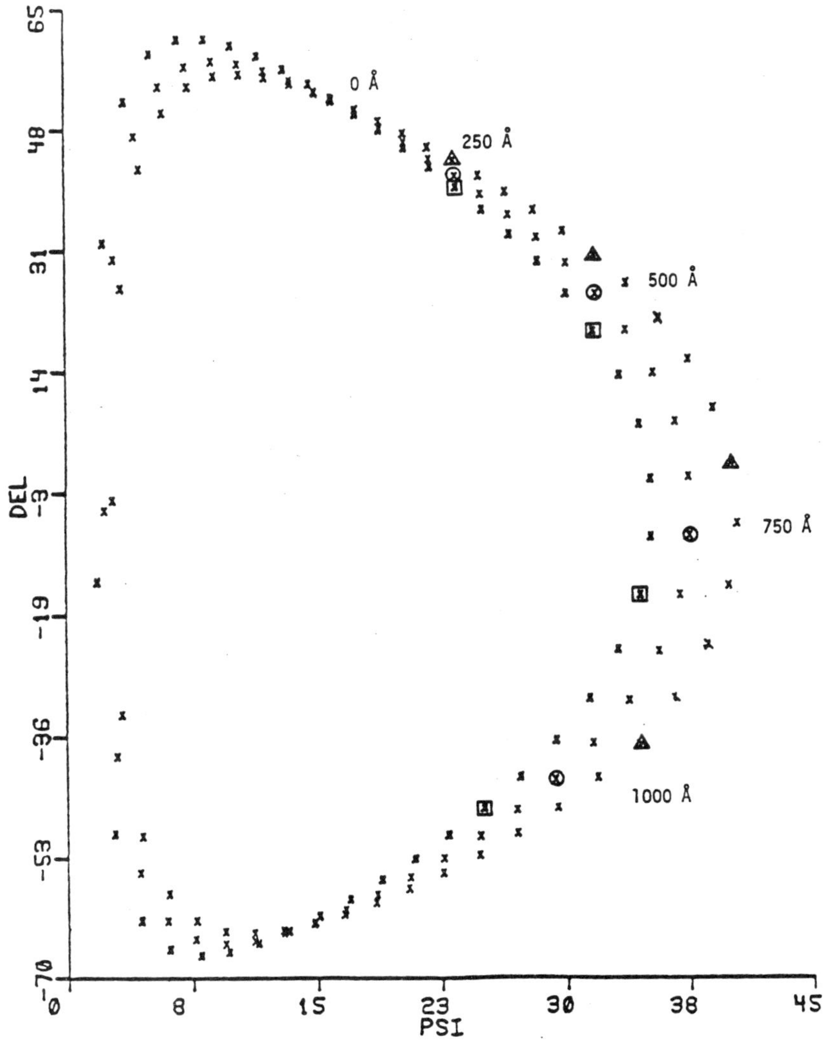

Figure III.2.3: Ellipsometric response function for films on glassy carbon with film refractive indices of 1.65 (△), 1.70 (O) and 1.75 (□). Other parameters are the same as Figures III.2.1 and III.2.2.

the angle of incidence will be changed. Errors due to small tilt angles may be corrected by computation [13,17].

Other errors, such as random errors, systematic errors (e.g. optical components not perfrectly aligned) and the resolution of the ellipsometer can greatly affect the accuracy of an ellipsometric measurement. The effects of these errors have recently been examined in detail by Reidling [18]. His work clearly demonstrates the benefits of computational modeling, which can aid greatly in evaluating and interpreting ellipsometric data.

References

1. Archer, R. J., J. Amer. Op. Soc., 52, 970 (1962).
2. ASTM Standard F 576-78, "Standard Method for Measurement of Insulator Thickness and Refractive Index on Silicon Substrates by Ellipsometry".
3. Stromberg, R. R., Passaglia, E., Eutas, D. J., "Application of Ellipsometry to the Study of Adsorption from Solution", in "Ellipsometry in the Measurement of Surfaces and Thin Films", E. Passagli, R. R. Stromberg, J. Kruger, Eds., NBS Miscellaneous Publication, 256, 1964.
4. Boerio, F. J. and Chen, S. L., Applied Spectroscopy, 33, 121 (1979).
5. Muller, R. H., Steiger, R. F., Somorjai, G. A. and Morabito, J. M., Surface Science, 16, 234 (1969).
6. Paile, W., Genshaw, M., and Bockris, J. M., J. Phys. Chem., 74, 4266 (1970).
7. Mabai, M. and Gottsfield, S., Surface Science, 96, 461 (1980).
8. Dynes, P. J. and Kaelble, D. H., Ref. 45, p. 167.
9. Azzam, R.M.A., nd Bashara, N. M., Ellipsometry and Polarized Light, North Holland PUbl. Co., New York, NY, 1977.
10. Zaininger, K. H. and Revez, A. G., RCA Review, 25, 85 (1964).
11. McCrackin, F. L., Passaglia, E., Stromberg, R. R. and Steinberg, H. L., J. Research of Nat'l Bureau of Standards - A. Phys. & Chem. Vol. 67A, 363 (1963).

12. Born, M. and Wolf, E., Principles of Optics, Pergamon Press, LTD, London, Ch. 13, 1965.

13. McCrackin, F. L., NBS Technical Note 479, issued April 1969.

14. McCrackin, F. L. and Colson, J. P., National Bureau of Standards Misc. Publ. 256, Sept. 15, 1964.

15. Hunderi, O., Sursface Science, 96, 1 (1980).

16. Ferstmaker, kC. A. and McCrackin, F. L., Surface Science, 16, 85 (1969).

17. McCrackin, F. L., J. Optical Soc. Amer., 60, 57 (1970).

18. Reidling, K., Thin Solid Films, 75, 355 (1981).

III.3 Experimental Section

Materials and Reagents

Solvents

Acetonitrile - Acetonitrile was UV "distilled in glass" grade (Burdick and Jackson Laboratories, Inc.). Solutions of acetonitrile and supporting-electrolytes were prepared within a few hours before their use in electrochemical experiments and discarded if more than one day old.

Methylene Chloride - Methylene chloride was "distilled in glass" grade (Burdick and Jackson). When used as a solvent for electrochemistry, the same procedures as those described for acetonitrile were used for preparation of supporting-electrolyte solutions and drying of the solvent.

Methanol - Methanol was spectroscopic grade (Fisher). It was used as a solvent for ultrasonically cleaning substrates.

Aqueous Solutions - All aqueous solutions were prepared using distilled, deionized water ($>1 M\Omega\ cm^{-1}$ resistance).

Supporting Electrolytes

Tetrabutylammonium perchlorate (TBAP) - TBAP (South-West Analytical) was electrometric grade. The purity was such that no further purification was necessary except for vacuum drying at $60^{\circ}C$ for 6h. Following drying, TBAP was stored in a vacuum desiccator.

Tetraethylammonium p-toluene sulfonate (TEApTs) - TEApTs was electrometric grade (South-West Analytical). TEApTs was dried *in vacuo* at 25°C for 6 h. and stored in a vacuum desiccator.

Tetrabutylammonium hexafluorophosphate ($TBAPF_6$) - $TBAPF_6$ was electrometric grade (South-West Analytical). $TBAPF_6$ was dried *in vacuo* at 25°C for 6 h. and stored in a vacuum desiccator.

Lithium Perchlorate ($LiClO_4$) - $LiClO_4$ was obtained as the tri-hydrate salt (G. Fredrich Smith Chemical Co.). It was used as received.

Reagents

Ruthenium tris-Bipyridyl bis-hexa fluorophosphate - ($Ru(bipy)_3(PF_6)_2$)- $Ru(bipy)_3(PF_6)_2$ was prepared by metathesis of $Ru(bipy)_3Cl_2 \cdot 6H_2O$ (G. Fredrich Smith Chemical Co.) was dissolved in 10 ml of water. This solution was added to an aqueous solution of HN_4PF_6 prepared by addition of 1 g of NH_4PF_6 to 10 ml of H_2O. Immediately upon mixing of the solutions, a red precipitate formed. The mixture was stirred for 10 minutes, after which the precipitate was separated by filtering the solution in a 10) pore size fritted funnel. The precipitate was washed with water several tims and then dried *in vacuo* for 12 h. at room temperature. The weight of the recovered precipitate was 0.230 grams, which gives a 99% yield for $Ru(bipy)_3(PF_6)_2$ based on the limiting reagent ($Ru(Bipy)_3Cl_2 \cdot 6H_2O$).

All of the reagents below were used as received from the sources listed.

Vinylferrocene - Polysciences

Hydroxymethyl ferrocene (HMF) - Strem Chemicals

9-10 Diphenylanthracene (DPA) - Aldrich

All other chemicals used not listed here were reagent grade or equivalent and were used as received.

Substrates

Choice of Substrates

An important goal of this part of the dissertation was to be able to accurately measure film thickness of PPVF film electrodes so that along with determination of electrochemical coverage, the concentration of electroactive sites in the polymer film could be calculated. For this reason, pryolytic gr aphite electrodes could not be used for this part of the study due to the roughness of these electrode surfaces which excludes the use of ellipsometric film thickness measurement. Following the guiding principle of seeking to immobilize potential electrocatalyst on readily available, and relatively inexpensive, conductive supports, PPVF films were deposited on vitreous carbon (glassy carbon) electrodes. Polishing procedures were developed by the author which made the glassy carbon electrodes suitable for use in ellipsometric studies. This allowed for both ellipsometric sutides and electrochemical stuides for PPVF films on the same substrate.

Silicon was the second major substrate used for PPVF depositions in this study. The main reasons for its use were: (1) it is a well suited substrate for optical studies and has been extensively characterized; (2) it has a very different refractive index from that of glassy carbon which aids in the interpretation of ellipsometric data by virtue of constancy of calculated PPVF film refractive index independent of substrate refractive index; and (3) since silicon is mostly transparent in the infrared, PPVF films deposited on silicon could be analyzed by infrared spectroscopy.

Polished platinum electrodes were also used as a substrate for PPVF films. Initially they were intended for use both in ellipsometric and electrochemical measurements. However, due to large deviations in the measured optical constants of bare polished platinum and a poor ellipsometer sensitivity function, film thickness measurements were not reliable. These substrates were useful, however, in determining if there was a substrate dependent PPVF deposition rate (compared to deposition on glassy carbon) by comparison of electrochemically determined coverage of PPVF on substrates prepared in the same plasma deposition. Also because of the wide potential window available with Pt substrates, examination of solution faradaic processes (with $E^{o'}$ far removed from that of PPVF) at PPVF film on Pt was possible.

• Glassy Carbon (GC) - Glassy carbon (Bitrecarb^R Vitreous carbon) was obtained from Fuororcarbon Co. (Anaheim, CA) in the form of plates measuring 13 x 13 x 0.32 cm. This source of glassy carbon was found to be superior to other sources (Tokai, Atomergic) in that the material was free of pits. The faces of the plates were ground flat and parallel with a diamond grinding wheel. Disks (0.95 cm diameter) were cut from the ground glassy carbon plates by electrode discharge machining in the University of Minnesota Physics department machine shop. All electrochemical and ellipsometric experiments with glassy carbon were conducted on these machined disks. The glassy carbon disks were polished to give an optically smooth surface by the procedure described below.

• Silicon - Doubly-polished single crystal silicon wafers (p-type) were generously donated by Dr. Robert Batdorf of Bell Laboratories (Reading, PA). Two different types of silicon wafers were obtained which were 0.07 and 0.20 cm thick with resistances of 180 Ω-cm^{-1} and 14Ω-cm^{-1} respectively. The 180Ω-cm^{-1} silicon wafers were cleaved to give 1.5 x 1.5 cm squares and were used primarily for infrared spectroscopic analysis of PPVF films deposited on these squares. The thicker silicon wafers (0.2 cm thick) were cleaved to give 0.9 x 0.9 cm squares and were used primarily for PPVF deposition rate studies (along with GC electrodes) as determined by ellipsometric thickness measurements. Prior to PPVF deposition on either of these types of silicon, these silicon squares were ultrasonically cleaned in methanol, dried, and etched in 49% hydrofluoric acid (ca 1 minute) to remove the thin natural oxide layer present on silicon. After etching, the silicon squares were rinsed with water, dried and immediately placed in a vacuum (either the plasma deposition reactor or the infrared spectrometer's vacuum chamber).

- Platinum (Pt) - Platinum plates (99.9% purity) were obtained from W.E. Mowrey Co. (St. Paul, MN) as 3 x 3 cm squares which were 0.04 cm thick. Disks (0.95 cm diameter) were cut from these plates for use in electrochemical experiments. These Pt disks were polished to a mirror finish following the same procedure for polishing of GC disks.

Polishing Procedure

Glassy carbon and platinum disks were ground flat with 600 grit Silicon - Carbide (SiC) paper (3M, St. Paul, MN) prior to polishing. The Si-C paper was secured on a flat glass plate by moistening the back of the paper. The grinding was done by hand on moistened Si-C paper using a figure eight pattern. Using a figure eight pattern as opposed to a circular pattern is important, since a circular pattern will preferentially grind the edges of the surfaces giving a contoured surface which is unacceptable for use in ellipsometric studies. Light hand pressure was used to avoid forming deep scratches in the substrate, which reduces the time spend on subsequent polishing stages needed to remove the surface damaged layer caused by grinding [1]. Following grinding, and between each polishing stage, substrates were ultrasonically twice cleaned in water, followed by ultrasonic cleaning in methanol. This cleaning step is very important, as it prevents carry-over of the grinding or polishing material to the next polishing stage.

Following grinding, rough and medium polishing of the substrates was achieved by rubbing the substrates sequentially over the following Si-C papers: 32, 23, 15, 8, 5, 3 and 1 micron particle size. (Moyco Industries, Philadelphia, PA). These Si-C papers were mounted and moistened as described above, and the same rubbing pattern was used. Each polishing stage required ca 2 minutes of rubbing. The substrates were very reflective at this point, but under examination by optical microscopy (magnification of 100x) very fine scratches were observed.

The final polishing stages consisted of hand polishing sequentially with diamond pastes (Buehler, Ltd., Lake Bluff, IL) which had mean diamond particle diameter sizes of 1.0μ and 0.25μ. The polishing was performed by loading the diamond paste onto a short napped, cotton cloth which was tightly fitted over a flat aluminum wheel. Again a figure eight pattern was used for the rubbing pattern. The substrates were ultrasonically cleaned as described above. After the final polishing, the glassy carbon disks were very reflective and free of scratches (as determined by optical microscopy at 100X magnification. Pt disks were also very reflective, having a mirror finish. However, examination by optical microscopy revealed some scratches and imperfections on the polished Pt disks.

The polished substrates were further characterized to establish the success of the above polishing procedure. A crucial observation, which determines whether the substrates may be used for ellipsometric measurements, is whether appreciable scattering of the incident beam from the ellipsometer light source occurs upon reflection from the substrate surface. The light source (He-Ne laser, λ = 6328) of the ellipsometer used in this work was equipped with microspot optics which gives a reflected image measuring 0.0025 x 0.0086 cm (at 70° angle of incidence) on the sample surface. The laser image should be practically invisible on optically smooth surfaces. This was the case observed for silicon substrates (which are known to be optically smooth), and polished GC surfaces. Appreciable scattering was observed for polished Pt surfaces, suggesting the surface was not as smooth as in the case of the GC surfaces. The difference in smoothness of GC and Pt surfaces even though the same polishing procedure was used is not unreasonable, since the mechanical properties of the two materials are quite different.

The necessity of using 0.25μ diamond for final polishing is evidenced by observable scattering when this stage is omitted (for GC disks) and differences in measured optical constants. The refractive indices of 1.0μ polished GC disks were 1.830 (1-0.402i), and for 0.25μ polished GC disks were 1.864 (1-0.395i). The reproducibility of the polishing procedure is evidenced by the reproducibility in measured optical constants. For 0.25μ diamond polished samples (25 separate disks were used for the measurements), the following optical constants were found:

$$n = 1.86 \pm 0.01, \quad k = 0.395 \pm 0.002$$

where n and k are the real and imaginary parts of the substrate refractive index. Much larger variations in measured refractive index were found for 0.25 μ diamond polished Pt disks (8 disks used for measurement):

$$n = 2.17 \pm 0.07, \quad k = 2.18 \pm 0.09$$

Because of these relatively large deviations in the measured refractive index of Pt, Pt disks were judged to be unsuitable for use as substrates in ellipsometric PPVF film thickness determinations.

Further evidence for the effectiveness of the 0.25 μ diamond polishing stage is given by examination of background current levels for 1.0 μ diamond polished and 0.25 μ diamond polished GC disks. The background current was reduced by _ca_ 25% for the 0.25) polished GC compared to the 1.0 μ polished GC as examined in 0.1M TBAP/acetonitrile by cyclic voltammetry. The lower current suggests a reduced surface area as would be expected for a smooth surface.

Examination of the polished GC disks by scanning electron microscopy at the maximum usable magnification (20,000x) revealed a completely featureless surface. The lack of observable features made it difficult to focus the microscope, focussing being achieved by virtue of finding an occasional dust particle on the surface.

The above results support the reliability of the polishing procedure developed here for polishing GC surfaces. Ross [2] had previously used alumina abrasives for polishing GC surfaces which has been more or less general procedure for polishing GC surfaces. This author found that even with the smallest alumina particle size (0.05 μ), a scratch free surface could not be obtained and greater degrees of light scattering were observed, which would lead to errors in the ellipsometric determinations. The exact reasons for the difference in the success of polishing using the two methods is not presently understood (possible differences in average particle size distributions or coefficient of friction between polishing compound and GC surface).

Characterization Apparatus and Procedures

Electrochemical Characterization Apparatus and Procedures

The cell used for all the electrochemical experiments was the general purpose cell described below.

Several features were designed into this cell that made its use facile and uncomplicated in terms of assembly. One feature which facilitates working with solid disk electrodes in terms of mounting and disassembly was the design of the Kel-F working electrode holder (E). The inside diameter of the Kel-F tube was 0.013 cm larger than the diameter of the PG electrodes (0.953) which allowed slipping the working electrode (PG) into the holder easily but securely. A teflon gasket was placed between the solution side of the working electrode and the retaining lip of the Kel-F holder. The brass electrode contact (Part C, both figures) pressed the electrode against the washer, via the threaded shaft, thus providing for a leak-tight seal and a reproducible geometric electrode area. The exposed area of the electrode was 0.24 cm^2. Since the Kel-F tube is semi-transparent, leakage would have been observed if it had occured.

Another design feature which allowed easy assembly of the entire cell was the use of Ace-Thread® fittings for the salt bridge and working electrode mounting. The entire cell, when disassembled, was ultrasonically cleaned in a water-isopropanol mixture, dried, and stored in a vacuum dessicator when not in use.

The Ace-Thread feature also allowed for variable positioning of the reference probe (salt bridge) and the working electrode holder. This afforded close-positioning of the reference probe to the working electrode, thereby minimizing uncompensated solution resistance. Typical distance between the probe tip and the working electrode was ca 2-3 mm. The probe end had a platinum wire-pyrex glass seal to provide electrolytic contact with the working medium. The probe (salt-bridge, Part H) contained a concentrated solution (typically 1.0 M) of the solvent-supporting electrolyte system used during the experiment. The use of the bridge prevented cross contamination of the reference electrode filling solution and the solution under study.

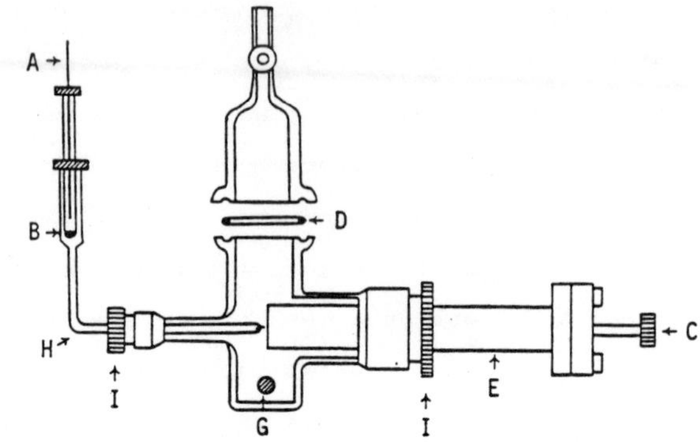

SIDE VIEW

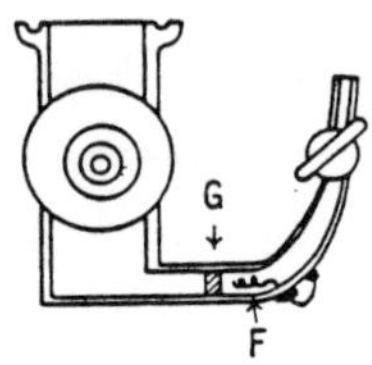

ANOTHER SIDE VIEW

Figure III.3.1: General Purpose Pyrex Cell

A) Ag/AgCl Wire

B) Cracked-Glass Bead Junction

C) Brass Electrode Contacting Screw Assembly

D) Silicone O-Ring

E) Kel-F Working Electrode Holder

F) Pt-Wire Auxilliary Electrode

G) Coarse Glass Frit Separating Working and Auxiliary Electrode Compartments

H) Salt Bridge - Probe

I) Teflon Bushing (Ace Thread®)

The potential reference used throughout this work was the Ag/AgCl half-cell in 1.0 M KCl (saturated with AgCl). These Ag/AgCl electrodes were prepared by the procedure described by Sawyer [8]. The Ag/AgCl wire electrode was housed in a pyrex tube with a soft glass/pyrex junction (cracked glass bead junction) to provide electrolytic contact with the salt bridge solution. The potential of these Ag/AgCl electrodes was very reproducible and stable over a period of 1-2 years, as long as the KCl solution was periodically renewed. When not in use, the Ag/AgCl cracked glass bead electrode was stored in a stoppered vial containing AgCl saturated 1.0 M KCl in the dark.

Oxygen purging of the electrolyte solution was achieved by inserting a long surgical needle through the top teflon stopcock and bubbling with high purity nitrogen. During experiments, the needle was moved up out of solution, but remained in the cell, passing a steady stream of nitrogen over the cell. All electrochemical experiments were performed with purged solutions.

When using methylene chloride as the electrochemical solvent, Kal-Rez^R o-rings (DuPont) replaced the silicon o-rings used for sealing.

All cells were de-oxygenated with nitrogen bubbling. An oxygen trap (Model RGP-250) obtained from Lab Clear, Inc. (Oxnard, CA) was used to remove trace oxygen from the nitrogen source. The purified nitrogen was then presaturated with the electrochemical solvent by passing it through a bubbler containing this solvent to avoid evaporation of the solvent in the electrochemical cell.

A Princeton Applied Research Model 173 potentiostat equipped with a Model 179 Digital Coulometer was employed for electrochemical measurements. Potential waveform programming was executed with a PAR 175 Universal Programmer. The 175 also provided trigger levels for triggering potential step perturbations and a digital waveform recorder for chronocoulometric and chronoamperometric experiments.

Cyclic voltammograms were recorded on a X-Y recorder (Model 815, Allen Datagraph). Current and charge transients were stored on a digital waveform recorder (Wavesaver, Epic Instruments). The waveform recorder could accept 1024 points at sampling intervals from 2 μs to DC. The output of the Wavesaver was recorded on a strip chart recorder (Heath-Schlumberger Model SR204) by clocking the data out at a rate of 20 Hz.

IR compensation was needed only in the case of electrochemical experiments employing methylene chloride as the solvent. This was established by examining the charging current response to a 40 mV p-p 1 kHz square wave potential perturbation as outlined in the PAR 179 manual. The iR compensation in methylene chloride was adjusted to ca 85% compensation. This was sufficient for the time scale and currents of the electrochemical experiments performed.

A test used to show that 85% compensation was sufficient was examination of peak potential separation while increasing iR compensation until no difference was observed in ΔE_p as a function of iR compensation. Greater compensation was not used because of possible loss of potential control due to potentiostat oscillation. In the case of PPVF film covered electrodes, loss of potential control would rapidly decompose the film, and hence critical compensation was avoided. IR compensation was not employed or needed when water or acetonitrile was employed as the solvent at unmodified or PPVF film covered electrodes (examined with varying thickness PPVF films) as established by both procedures described above. Initially, surface coverage determinations and thermodynamic studies were attempted by cyclic voltcoulometry. However, due to significant background currents due to some background reaction(s), the double potential step chronocoulometric method described in a later section was used for surface coverage determination and calculating ratios of oxidized/reduced ferrocene in PPVF films at 50 mV intervals. The hold time for the anodic step was 2 min. in all cases. To establish that this was sufficient for eqilibration, the hold time was increased and the calculated surface coverage was found to be independent of hold time greater than 2 minutes. In reporting the data to follow, Q_T is defined as the charge due to total conversion of ferrocene to ferricenium, and Q_a is the charge due to conversion of ferrocene to ferricenium at a potential (E).

All potentials are measured with respect to the aqueous 1.0 M KCl Ag/AgCl reference electrode. The PAR 173 can be configured as potentiometer or electrometer. This feature was used to measure the potential difference between extracted PPVF film electrodes and Ag/AgCl reference electrode prior to other electrochemical experiments. This potential difference is defined here as the rest potential. Rest potential measurements were used to examine the amounts of ferricencium initially present in the PPVF films.

Ellipsometric Apparatus and Procedures

The ellipsometer employed for measuring optical constants and film thickness was a Gaertner Model L117 Ellipsometer housed in the University of Minnesota Microelectronics laboratory. The ellipsometer was equipped with a viewing microscope, primarily used for sample positioning and focusing (Model L117UM), microspot optics which allowed examinations of small areas (25 x 86 μ), and the light source was a He-Ne laser (λ = 6328). The angle of incidence could be fixed at 30°, 50° and 70°, but in general only the 70° angle was employed for ellipsometric measurements. In general, three measurements were made on each sample and the resultant calculated optical constants or film thickness were then averaged. To examine film thickness variations, the sample was moved (by x-y manipulator stage) across the incident beam and the points of largest variation in polarizer and analyzer readings were recorded in order to establish the extent of non-uniformity in film thickness.

Optical constants and film thickness were calculated using the McCrackin program [3]. The program was obtained from Frank McCrackin (Characterization and Standards Group, Polymer Science and Standards Division, National Bureau of Standards, Washington, D.C.) in the form of punched cards and loaded onto a VAX 11/780 computer (Digital Equipment Corp.) housed in the University of Minnesota Chemistry Department. Only minor program modifications were made, mainly for the purpose of formatting input and output statements. To allow for calculation of any thickness period (τ), two statements in the subroutine CALDX [3] were changed as follows [180]:

1) T_5 = CLOG (Y1)

2) T_6 = CLOG (Y2)

Modified statements

1)' T_5 = CLOG (Y1) - (LOOP-1)*(0.,6.203185)

2)' T_6 = CLOG (Y2) - (LOOP-1)*(0.,6.203185)

The value of LOOP(m) was made an input parameter in the computer program.

For all of the PPVF film thickness measurements reported, no assumptions were made as to the film refractive index. The film refractive index and film thickness were calculated simultaneously. This was possible by taking advantage of the different ellipsometric sensitivity functions of glassy carbon and silicon substrates. In general, for films greater than 1000 Å thick (or > 1000 Å+mτ), silicon substrates were used for accurate film refractive index calculations. For films less than 400 Å thick (or >400 + mτ), glassy carbon substrates were used for assignment of film refractive index. For films between 400 Å and 1000 Å, where both substrates give good sensitivity, calculation of PPVF film refractive index (nf) and thickness on both substrates were the same within experimental error (95% confidence limits). The confidence limits were established by inputting the uncertainty in Δ and ψ (0.4 and 0.2 degrees respectively [4] and/or the uncertainty in the substrate refractive index (whichever causes the largest deviation) into the McCrackin Program. The substrate which gave the smallest confidence interval (at a given film thickness) for the calculated film refractive index was used to assign the film refractive index. In general, the PPVF film refractive index confidence limits were equal to nf ±0.015. The exception was for films 200 Å thick or less (or <200 Å + mτ), where the confidence limits were nf ±0.04.

Other Apparatus

PPVF films deposited on silicon substrates were characterized by infrared spectroscopy using a Fourier - transform infrared spectrometer (Digilab model FTS-20M). This spectrometer was equipped with a dedicated computer (Data General, Nova 3) for calculating spectra and data handling. All spectra reported were taken *in vaccuo* with the instrument resolution set at 8 cm^{-1}. Spectra of PPVF deposited on silicon were ratioed against bare silicon wafers to give the spectra of PPVF films without spectral contributions from the silicon substrate. Spectra of authentic polyvinylferrocene (PVF) were acquired from samples prepared by pipeting of solution of PVF in methylene chloride onto a KBr wafer and then evaporating off the methylene chloride. The PVF was generously donated by Professor Allen Bard (University of Texas, Austin, TX).

XPS characterization of PPVF films was conducted on a model 555 Electron Spectrometer (Physical Electronics) equipped with a MgK_α -X-ray source. This instrument was housed in the Surface Analysis Center at the University of Minnesota. The model 555 was also equipped with a Model 5700 computer operating system (Physical Electronics) for total system control during analysis and data handling.

Scanning electron microscopy on bare and PPVF film covered substrates was conducted on a Cambridge electron microscope (Stereoscan model) equipped with an x-ray fluorescence detector (Kevex X-ray analyzer). This instrument was housed in the University of Minnesota Electron Microscopy Center.

Plasma Deposition Apparatus and Procedure for PPVF Deposition

A substantially different plasma deposition apparatus was designed for deposition of PPVF films in order to acquire more control over the polymerization process and to better monitor the system variables. Several improvements over the original prototype plasma reactor that were deemed necessary include: the containment of the plasma reactor in a thermally well regulated oven; the containment of the monomer supply in a separate oven, so that the monomer supply temperature may be independently controlled; provisions for accurate absolute pressure measurement; the use of high vacuum components in reactor construction and the use of a diffusion pump, so that lower system base pressure could be achieved; the ability to achieve steady state plasma reaction pressures; and the incorporation of a larger diameter reactor chamber. This last requirement was necessary to minimize "wall effects" in the plasma reactor which cuase non-uniform deposition (in terms of film thickness) for samples whose lateral dimensions approached the diameter of the reaction vessel. These wall effects were severe in PPVF depositions on 1.5 centimeter square silicon wafers using the prototype plasma reactor.

The plasma reactor with some of the associated hardware is shown schematically in Figure III.3.2. The diameter of the reactor (constructed of pyrex glass) was 8.5 cm and the total length of the reactor (parts T and B) measured 31 cm giving a reactor volume of 1800 cm^3. A fluoroelastomer o-ring provided for a high vacuum seal between parts T and B. The reactor top (part T) was secured by sue of a ring clamp (not shown) which allowed for easy removal of the reactor for cleaning purposes, yet gave very reproducible positioning (relative to the excitation coil)

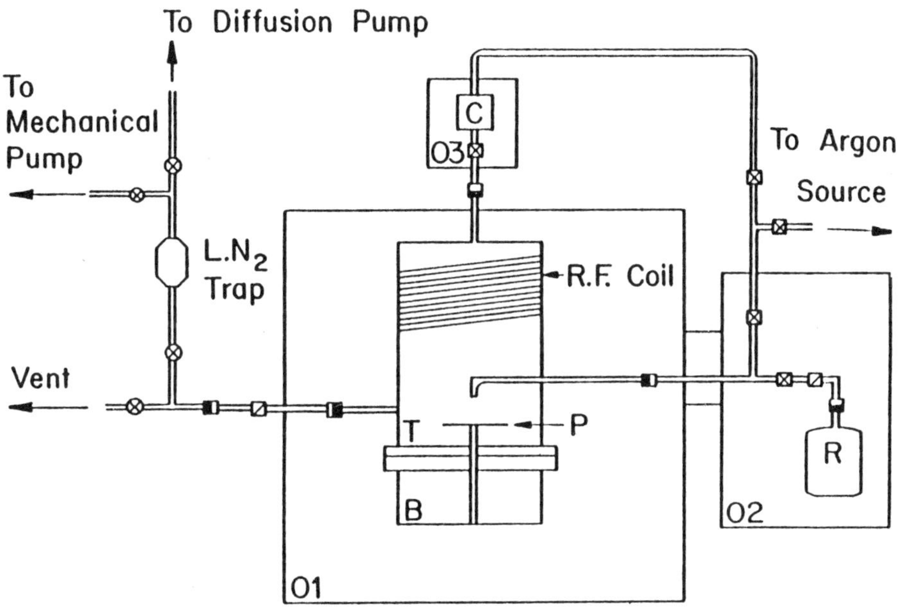

Figure III.3.2: Plasma Deposition Apparatus

C = Capacitance manometer

T = Top of plasma reactor

B = Bottom of plasma reactor

P = Sample pedestal

R = Monomer reservoir

01 = Plasma reactor oven

02 = Monomer reservoir oven

03 = Capacitance manometer oven

◨ = Ultra-Torr coupling, shaded side connects to glass, unshaded side connects to stainless steel tubing or stainless steel bellows

⊠ = Stainless steel bellows shut off valves

⊗ = Glass/teflon high vacuum valves

◿ = Stainless steel bellows regulating/shut off valves

upon reassembly. The reactor bottom had affixed to it a coaxially located glass pedestal (part P) onto which were placed substrates for PPVF film deposition. The diameter of the pedestal was 3.8 cm, hence substrates placed on the pedestal were well removed from the reactor walls.

Connections from the reactor to the monomer reservoir, vacuum line, and capacitance monometer were made with stainless steel Ultra-Torr^{R} fittings (Cajon Vacuum Products) which were welded to stainless steel bellows. The use of the bellows made disassembly of the reactor possible without need for realighment after reassembly.

All of the valves employed in the three ovens were stainless steel ultra high vacuum bellows valves (Nupro Company). The valve handles were modified so that they extended through the oven walls permitting the opening and closing of these Valves externally.

All of the tubing and connections were constructed of pyrex glass or stainless steel as depicted in Figure III.3.2. The monomer reservoir (R), constructed of pyrex glass, had a volume of 200 cm^3. All glass vacuum valves were high vacuum valves with teflon barrels and fluoelastomer o-rings (Ace Glass).

The coil used to couple the output of the RF generator to the reactor was constructed of ten turns of 3.18 mm diameter copper tubing and was 9.5 cm in diameter and 5.6 cm long, giving an inductance of ca 11 μ H. The impedance of the discharge was matched to the RF generator by the use of a -type matching network (Tegal Corporation). The length of the leads from matching network (located was outside the reactor oven) to the excitation coil was minimized to decrease the parasitic capacitance associated with the separation of these devices. The RF generator, which operated at a fixed frequency of 13.56 MHz, capable of delivering up to 300 watts RF power, was obtained from Tegal Corporation (Model 300P). The power absorbed by the discharge was measured by Model 43 "Thruline"-RF Directional wattmeter using a 50 watt probe (Bird Electronic Corporation). The meter allowed for measuring the forward and reflected power to 0.3 watts resolution.

Pressure was monitored with a capacitance monometer (Baratron differential gauge, Model 221 AH-10, 10 torr head) obtained from MKS instruments. This gauge was operable up to 100°C and was equipped with signal conditioning electronics (separated from the gauge by 2.4 cm cable) giving an output of 1 mV/mtorr. The output of the gauge was read by a digital voltmeter. The guaranteed accuracy of the gauge is rated to 0.5% (±temperature coefficients) of the reading between 10 torr and 10 mtorr. Control of the gauge temperatures to ±1°C extended the useful range down to <1 mtorr, although the accuracy between 10 and 1 mtorr was not specified. The accuracy in this range is assumed to be no better than ±1 mtorr.

In operation of the gauge, it was used as an absolute pressure meter. The reference side was evacuated to below the resolution of the gauge and isolated during pressure measurements. In general, this procedure was sufficient in terms of pressure monitoring accuracy for most experiments. However, since the reference side could not be continuously evacuated during a pressure measurement (due to the design of the plasma deposition system) some drift occured in the reading at long measurement (ca 1-3 mtorr/hour). This caused some uncertainty in pressure monitoring which, not initially expected, was relatively important for very long time scale PPVF deposition experiments vide post).

Mechanical pumping was achieved with a Directorr pump (Model 8805, Sargeant Welch), which served both for rough pumping of the vacuum system and as a forepump for the diffusion pump. The diffusion pump was a water cooled, three stage oil diffusion pump built by the University of Minnesota glass shop.

The monomer reservoir oven and the plasma reactor oven were commercially available convection ovens, and the oven that housed the capacitance monometer was built in house. Accurate temperature control of the three ovens was achieved with the use of solid state electronic temperature controllers (VSI Thermistemp^{R} Model 63RC, Yellow Spring Instrument Company) in conjunction with thermistor probes (YSI 621 thermistor probe). The probes were placed close to the heating elements in order to minimize the controller/oven time constant. Oven temperatures could be controlled to within ±1°C with these controllers. Oven temperatures were measured with copper-constantan thermocouples and monitored on a digital temperature indicator (Model 199, Omega Engineering). The accuracy of temperature measurement was ca 1°C. The plasma reactor oven was modified to include a viewing window to allow observation of the plasma during depositions.

Plasma Deposition Procedure

The plasma was "tuned" by igniting a 200 mtorr Argon discharge while minimizing the reflected power with the matching network. The coupling efficiency for the powers used in this part of the dissertation (5-9 watts net power) was ca 60%, where the coupling efficiency is defined as the net power divided by the forward power. Once tuned, the matching network was left fixed for all the work described in here. While sustaining an argon plasma discharge, the pressure was lowered by pumping while examining the coupling efficiency. It remained constant to pressures of down to 1 mtorr, and in addition the net power remained constant. The plasma filled the entire reaction chamber at 200 mtorr Argon pressure, but inhomogeneities in luminous intensity were observed. However, as the pressure fell, the plasma intensity became more homogeneous at pressures between 60 and 1 mtorr except for a pronounced wall effect in the coil region. (The walls of the reactor near the coil appeared dark.) The plasma intensity was very homogeneous in the area around the sample pedestal. No additional excitation source was required to ignite the plasma, and the plasma could be sustained with 2 watts net power, but 3 watts were needed to ignite the plasma. Argon plasmas were used prior to deposition experiments (in the absence of substrates) to set the initial power for the deposition.

The monomer reservoir flask was loaded with 0.1 grams of vinylferrocene prior to each deposition. The flask was cooled and maintained at 0°C with an ice-water bath and evacuated with the mechanical pump to below the resolution of the pressure gauge (<1 mtorr). Additional mechanical pumping on the vinylferrocene reservoir was performed for 1 h, followed by 0.25 h of pumping with the diffusion pump. After diffusion pumping, some vinylferrocene was observed to have collected in the liquid nitrogen trap, and henceforth, this was used as an indicator of sufficient pumping time (generally 15-20 min. with the diffusion pump). At this point, the vinylferrocene reservoir was isolated from the main chamber and allowed to warm to room temperature.

Substrates were then loaded into the reactor and mechanically pumped on for a minimum of 8 h with the plasma reactor oven temperature controlled at 55°C. The capacitance monometer oven was maintained at 65°C. The higher temperature in the capacitance monometer oven was used to prevent condensation of vinylferrocene onto the capacitor diaphram, which would lead to faulty pressure readings. Prior to deposition, the plasma reactor was evacuated for a minimum of 2 h with the diffusion pump. Both reference and sample sides of the capacitance monometer were evacuated during this time. The reference side of the gauge was then isolated immediately prior to deposition.

The monomer reservoir oven was turned on 1 h before deposition initiation to allow sufficient time for temperature equilibration. Initially it was thought that the monomer reservoir oven should be maintained at a higher temperature than the plasma reactor oven since deposition pressures in the prototype reactor fell rapidly, presumably due to exhaustive consumption of vinylferrocene monomer. Depositions on 3.81 cm diameter silicon wafers were attempted (at 5 watts net power) with the reservoir oven maintained at 65°C. In these experiments, the resultant PPVF films were non-uniform, in terms of thickness and reflectivity. During the plasma deposition, streaming could be seen in the plasma reactor, suggesting particle formation in the gas phase [5], giving rise to the non-reflective coating.

In light of the above, PPVF depositions on silicon wafers were attempted with the temperature of the monomer reservoir adjusted to 45°C, the other conditions being the same. In this case, the resultant PPVF films were very reflective, but there was substantial variation in film thickness across the wafer. The deposition was then attempted with the temperature of the monomer reservoir adjusted to 50°C. In this case uniform thickness and reflective films were achieved. It was noted, however, that the pressure varied during the deposition.

Even though the vinylferrocene reservoir temperature sutdies were by no means exhaustive, it was felt that a tamperature setting of 50°C allowed for formation of films which were suitable for the intended studies. The approach taken to achieve a steady state plasma pressure was to vary the RF power in order to maintain the pressure. The effect of the steady state pressure chosen will be discussed below. The final procedure which was developed for deppositing PPVF films was as follows.

The monomer reservoir oven was set at 50°C. Immediately before admitting vinylferrocene into the plasma reactor, the reactor was isolated from the vacuum pump. Vinylferrocene was then admitted into the deposition chamber and the pressure allowed to rise to 35 mtorr prior to igniting the plasma. This typically took 2 min., and this was a good indication that no leaks were present in the monomer reservoir connections. The plasma was then ignited using a net power of 7 watts. The plasma would achieve full luminosity within 10 to 20 seconds, and the pressure would fall to less than 10 mtorr witin ca 30 s. At this point, if necessary, the net power was manually adjusted to maintain the pressure between 6 and 8 mtorr throughout the deposition. The net powers used were between 5 and 9 watts. After deposition, the PPVF covered substrates were left in the reactor with ca 20 mtorr vinylferrocene vapor for 0.5 hour. The reactor was then pumped down to <1 mtorr, at which point the reactor oven turned off. After cooling to room temperature, the film covered substrates were removed from the reactor. The total deposition time defined here was the time from when the RF power was turned on to when the RF was turned off. Deposition times ranged from 5 to 70 min.

Except where specified, the resultant PPVF films were extracted in acetonitrile (Soxhlet) for 8 h prior to characterization.

Evaluation of Reactor Performance and Deposition Procedure

The intitial test for establishing suitable conditions for plasma deposition of vinylferrocene was the demonstration that reflective films of uniform thickness could be generated. In the initial evaluation, the RF power was maintained at 5 watts net power for all depositions while varying the temperature of the vinylferrocene reservoir oven. Maintaining the reservoir oven at 50°C afforded the generation of reflective and highly uniform thick films. For example using these conditions for a 15 minute PPVF deposition on a 3.8 cm diameter silicon wafer, an average thickness of 1060 ±37Å was found. The thickness variations are not random, rather the variations are continuous with the film getting thicker towards the edge of the wafer. The thickness in the center of the wafer was 1040 Å, and the thickness at the very edge was 1119 Å. The average thickness was calculated by taking the average value of 5 points spaced at 0.75 cm intervals along each axis defining the plane of the wafer. The thickness variations were most extreme towards the edge of the wafer. Although the variation in thickness was still significant, the variation in thickness on the exposed area of substrates used for electrochemical experiments (0.24 cm^2) would be minimal (less than 20 Å variation in thickness across the face of the electrode for ca 1000 Å thick film).

Judging the deposition procedure elaborated on in the previous chapter to be suitable, the procedure was further evaluated by replicate experiments establishing the deposition rate as a function of deposition time (at constant RF power). The thicknesses of the as-deposited PPVF films were linear with deposition time and had constant refractive index, but large variations in the thickness and refractive index of the extracted films were found for replicate depositions (constant deposition time) and for depositions as a function of time. The point of greatest concern was the variation in the refractive index of the extracted films (ca 1.4-1.7), since the refractive index of the film is related to the density of the film [3]. It was deemed necessary to have films of constant refractive index (hence constant density) for the studies to be described later.

The source of the irreproducibility of the extracted film refractive index was attributed to the variation in pressure during the deposition using the above constant power procedure. Ross [2] had reported similar variations in extracted film refractive index for PPVF films deposited on silicon using the prototype reactor.

The pressure generally increased during a given deposition, ranging from ca 5 to 25 mtorr in a 30 minute deposition. The increase in pressure was presumed to be due to the kinetics of the polymerization being slower than the rate of delivery of monomer. As a consequence of the manner in which the monomer delivery rate was controlled (temperature of monomer reservoir) it would be difficult to adjust the monomer delivery rate rapidly due to the time constant associated with temperature control. An alternative method for achieving steady state pressure is to vary the RF power. The time constant associated with this control parameter is much smaller. The powers used to achieve steady state plasma pressure were 7 ±2 watts.

The effect of maintaining steady state pressure was examined and the results are given in Table III.3.1.

Table III.3.1 Effect of Steady State Pressure on PPVF Depositions on Glassy Carbon Electrodes (30 Minute Deposition)

	Unextracted		Extracted			
Pressure mtorr	d(u)[a] Å	nf	d(e)[a] Å	nf	d(e)/d(u) %	M[b] Moles/Liter
15 ± 2	2700±48	1.70±0.01[c]	778±25	1.52±0.01	29	2.59
12 ± 2	2723±44	1.71±0.01[c]	840±17	1.61±0.01	31	3.78
7 ± 1	2380±50	1.71±0.025[d]	1394±15	1.71±0.005	59	4.43

a) d(u) is the film thickness of the as-deposited film, d(e) is the film thickness after extraction in acetonitrile.

b) M is calculated by dividing the electrochemical surface coverage (Q), coulometrically determined in 0.1 M TBAP/CH_3CN, by the extracted film thickness (d).

c) In this case, the film thickness variation was significantly larger than the uncertainty in determining d, and the deviation reflects nonuniform film thickness.

d) In this case the uncertainty in determining d is larger than the thickness variation. Δ and ψ measured at several points on the film covered surface were invariant.

Even though films deposited at higher pressure were initially thicker, the percentage of the film retained after extraction increased at lower pressure. The refractive index and the density of the extracted films (as evidenced by the concentration of electroactive ferrocene centers present in the film) increased as the steady state pressure was decreased. Films formed at the lowest steady state pressure studied were considerably more uniform in thickness across the surface of the sample. The films formed at lower pressure (higher refractive index) were found to be more stable in electrochemical experiments. For these reasons, the low steady state pressure conditions (7±1 mtorr) were used to generate PPVF films for the intended studies.

The results for PPVF depositions on glassy carbon, silicon and Pt substrates under these conditions as a function of total deposition time are summarized in Table III.3.2. The extracted and unextracted film thicknesses as a function of deposition time are shown in Figure III.3.3.. The linearity of the "as deposited" film thickness vs. deposition time plot is very good; the regression line gives a deposition rate of 80 Å/min and an intercept of -38 Å (correlation coefficient was 0.9998). The negative intercept is most probably a consequence of the finite time required for the initiated plasma to reach full intensity, this time being typically 20-30 seconds. The plot of extracted film thickness vs. deposition time shows considerably more scatter. Examination of the ratio [d(e)/d(u)] of the extracted to unextracted film thickness in Table II shows that this ratio was

Table III.3.2: Results for PPVF Depositions at a Steady State Pressure of 7 ±1 mtorr.

Deposition time (minutes)	Number of Depositions	Number of Samples	Unextracted $d(u)^a$ Å	Unextracted nf	Extracted $d(e)^a$ Å	Extracted nf	$\frac{d(e)}{d(u)}$ %	$\underline{M}^b$ Moles/Liter
5	1	3	395±12	1.69±0.015	218±32	1.70±0.04	55±5	4.4±0.2[c]
10	1	4	765±21	1.70±0.01	469±13	1.68±0.015	61±0.6	4.27±0.1[c]
15	2	10	1177±45	1.70±0.005	770±23	1.70±0.01	65±2	4.44±0.06[d]
30	1	4	2287±105	1.71±0.025	1328±63	1.71±0.005	58±2	4.48±0.06[e]
70	1	2	5596±326	1.72±0.01	4378±250	1.72±0.015	78±1	4.04±0.05[c]

[a] The deviation presented is either the uncertainty in determining d or the variation in thickness among samples. The thickness variation for a single sample was in all cases less than the uncertainty in determining d. Except for the 5 minute deposition, the thickness deviation was due to difference in thickness between samples.

[b] Concentration determined by dividing coulometrically determined surface coverage of a given sample (in 0.1 $\underline{M}$ TBAP/CH_3CN) by the respective sample film thickness.

[c] 1 sample used for determination

[d] 3 samples used for determination

[e] 2 samples used for determination

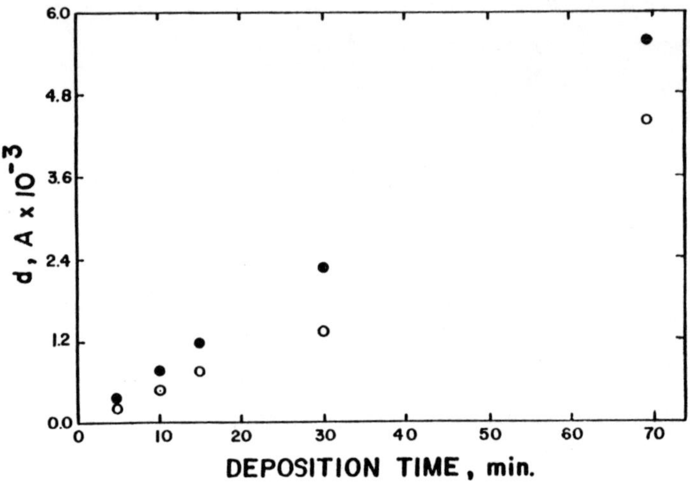

Figure III.3.3: PPVF film thickness as a function of deposition time.

● unextracted film thickness

○ extracted film thickness

not constant as a function of deposition time even though the pressure was controlled between 6 and 8 mtorr. The greatest variation was found for the 70 minute deposition. That variation in d(e)/d(u) occurred is not surprizing considering the results for films deposited at different steady state pressures. Here it is seen that d(e)/d(u) is a very strong function of the steady state pressure. The degree of pressure control required approached the usable resolution of the capacitance monometer. This was especially problematic for the 70 minute deposition, since drift in the monometer made the uncertainty in pressure measurement equal to the steady state pressure window (2 mtorr). Excluding the point at 70 minutes, the linear regression slope for d(e) vs. deposition time was 44 Å /min (intercept = 36 Å , correlation coefficient = 0.994).

Although d(e)/d(u) varied, the properties of the films were found to be very reproducible. The refractive index (nf) of all the films made (at 6-8 mtorr) was nearly invariant, the average refractive index of the unextracted films was 1.70±0.01 and the average refractive index of the extracted films was 1.70 ±0.02. That the refractive index was constant throughout the film is evidenced by the constancy of the electrochemically determined concentration of ferrocene in the PPVF films as a function of thickness. The average concentration was 4.40±0.09M, excluding the 70 minute deposition. A lower concentration was found here is attributed to rupture of the film during electrochemical characterization, whereby parts of the film might have been electrochemically "silent" on the time scale of the electrochemical measurements made (vide post). The constancy of the film refractive index and concentration of electroactive sites allows valid comparisons to be made between films of different thickness studied in various solvent-supporting electrolytes for evaluation of PPVF film thermodynamic properties.

For the substrates studied here (silicon, platinum, and glassy carbon), the PPVF deposition rate was found to be independent of the substrate used. In the case of comparisons between glassy carbon and silicon, this was established by ellipsometrically determined film thickness. Electrochemical coverage determinations were used for comparisons between glassy carbon and platinum. In contrast Nowak et al. [6] reported PPVF deposition rates which were 4-10X lower on glassy carbon vs. platinum (determined by electrochemical coverage). Their results were for PPVF films prepared in a plasma reactor which contained a charge of solid vinylferrocene while using 200 mtorr Argon as a diluent in the plasma discharge. They attributed this rather large difference in deposition rate to inductive heating of the substrate and/or quenching of the plasma by the substrate and these processes were postulated to be strongly affected by the substrate material. However, their results might have been a consequence of different mounting procedures used for glassy carbon and platinum substrates, placing these substrates at different positions relative to the solid monomer. In the same paper Nowak _et_ _al_. report a strong positional dependence for deposition rates [6] and hence this may have accounted for the variation in deposition rates they observed on glassy carbon and platinum. Quenching of the plasma by the substrate might also have had some effect at the pressure regime used by these investigators. The space charge (dark space) region was observed to be more pronounced for argon plasmas at higher pressures examined in the reactor shown in Figure III.3.2. The dark space decreased as the Argon pressure was reduced, becoming practically indiscernable at very low pressure (1-20 mtorr). Since PPVF depositions in this dissertation were conducted in this pressure range, plasma quenching effects are minimized. The axial symmetry of the reactor used in this work removes the ambiguity of positionally dependent deposition rates. With position and quenching effects removed, expectation of substrate independent deposition rates is reasonable considering that once the first "equivalent monolayer" is deposited, the underlying substrate surface propertie which might influence deposition rate are effectively "masked" from the plasma.

Other PPVF film properties investigated were found to be independent of substrate. PPVF film refractive index (extracted or unextracted) deposited on glassy carbon or silicon was calculated and found to be the same within experimental error for films whose thickness were in the sensitive regime of the ellipsometric response function of both substrates. The consequences of this observation are two-fold: (1) It supports that films deposited on different substrates are equivalent; (2) PPVF films are nonabsorbing [7] (k=0 or its value is immeasurable for the thickness of films studied here).

The amount of material extracted from "as deposited" films was independent of substrate and the length of time between deposition extraction; the d(e)/d(u) ratio was the same for films extracted within a few hours after deposition or one month after deposition. This implies that no further significant polymerization or crosslinking occurs after removing the substrates from the plasma chamber, exposing them to air.

The electrochemical behavior of PPVF films deposited on glassy carbon, platinum and pyrolytic graphite was equivalent both in terms of kinetic and thermodynamic behavior.

The most important consequence of thse findings is that examination of a particular property of PPVF films on a single substrate can be made without having to assume that this property will not be different on another substrate.

References

1. Samuels, L. E., Metallographic Polishing by Mechanical Methods, Sir Isaac Pittman and Sons, Ltd., Publ., 1971.
2. Ross, M. R., Ph.D. Dissertation, March 1981, University of Minnesota, Minneapolis, MN.
3. McCrackin, F. L., NBS Technical Note 479, issued April 1969.
4. Reidling, K., Thin Solid Films, 75, 355 (1981).
5. Kobayashi, H., Bell, A. T. and Shen, M., Macromol., 7, 277 (1974).
6. Nowak, R. J., Schultz, F., Umana, M. and Murray, R. W., Anal. Chem. 52, 315 (1980).
7. McCrackin, F. L. and Colson, J. P., National Bureau of Standards Misc. Publ. 256, Sept. 15, 1964.
8. Sawyer, D. T. and Roberts, Jr., J. L. Experimental Electrochemistry for Chemists, John Wiley and Sons, New York, 1974, p. 40.

III.4 Characterization of PPVF Films

Surface Characterization

Figure III.4.1 shows the XPS survey spectrum of a 752 Å film (extracted) deposited on glassy carbon examined within a day after its preparation. The spectral features are qualitatively similar to PPVF films deposited on pyrolytic graphite electrodes [1] and silicon substrates [2] using the prototype plasma deposition apparatus. As had been previously observed, a relatively high concentration of oxygen was present on the sample surface.

The high resolution XPS spectra taken in the Fe ($2p_{3/2}$), O(1s), and C(1s) regions are shown in Figures III.4.2, III.4.3, III.4.4, respectively. The elemental ratios measured from the high resolution spectra were: O/C = 0.266 and Fe/C = 0.076 (both valences of iron included). The Fe/C ratio is close to that expected for authentic polyvinylferrocene (Fe/C = 0.083). However, if the surface composition is presented in terms of fractional composition, the PPVF surface composition which was 74% C, 20% O, 6% Fe is clearly different from polyvinylferrocene (8% Fe, 92% C). Originally it had been expected that the deposition of vinylferrocene using a more sophisticated plasma reactor vacuum system (which could achieve lower base pressures and had lower leak rates) might afford PPVF films with reduced oxygen content or even oxygen free. This expectation was not found (compare elemental composition to that for films formed in prototype reactor, ref [1]). Without the ability to measure the partial

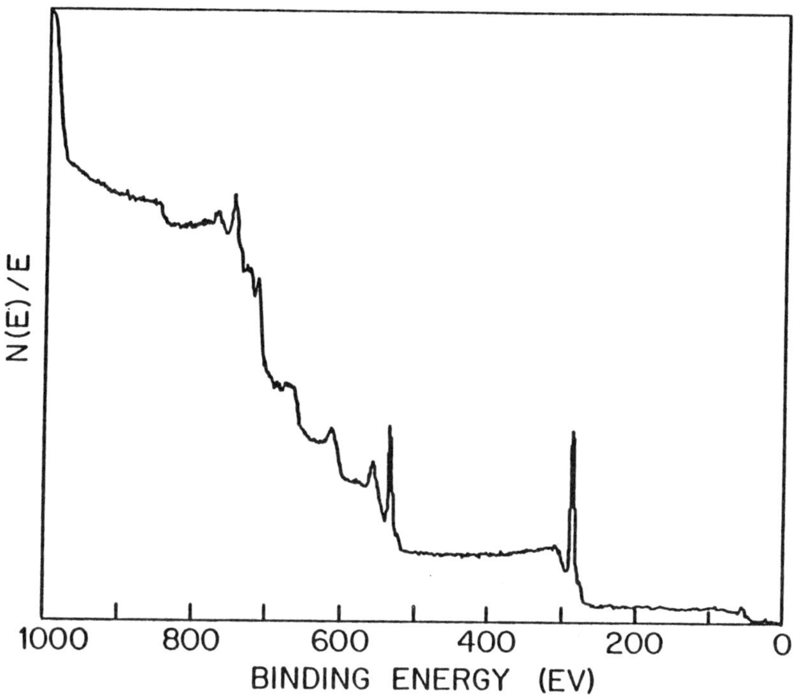

Figure III.4.1: Low resolution XPS survey spectrum of a 752 Å film of PPVF on glassy carbon. Pass energy = 200 eV.

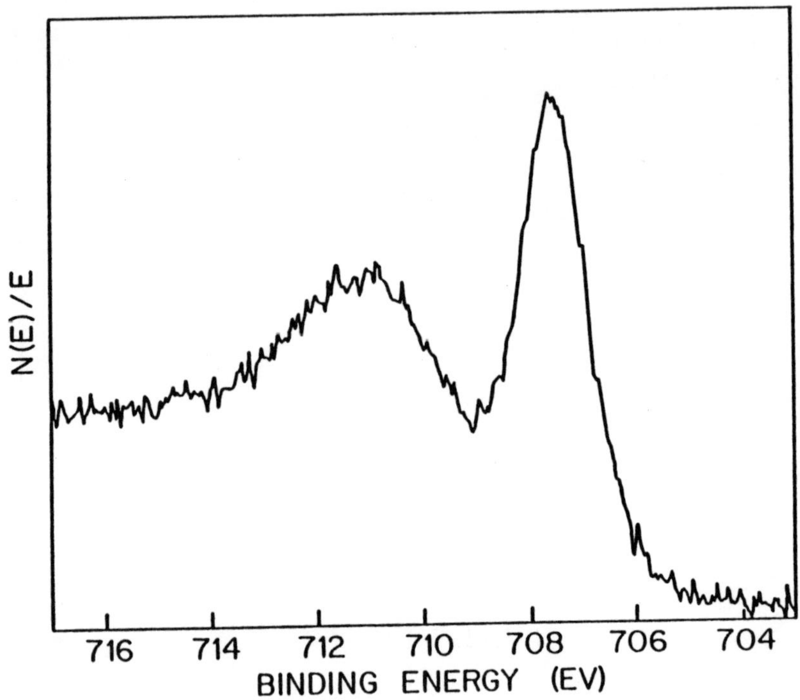

Figure III.4.2: High resolution XPS spectrum of a 752 Å film of PPVF on glassy carbon showing the Fe ($2p_{3/2}$) region. Pass energy = 25 eV.

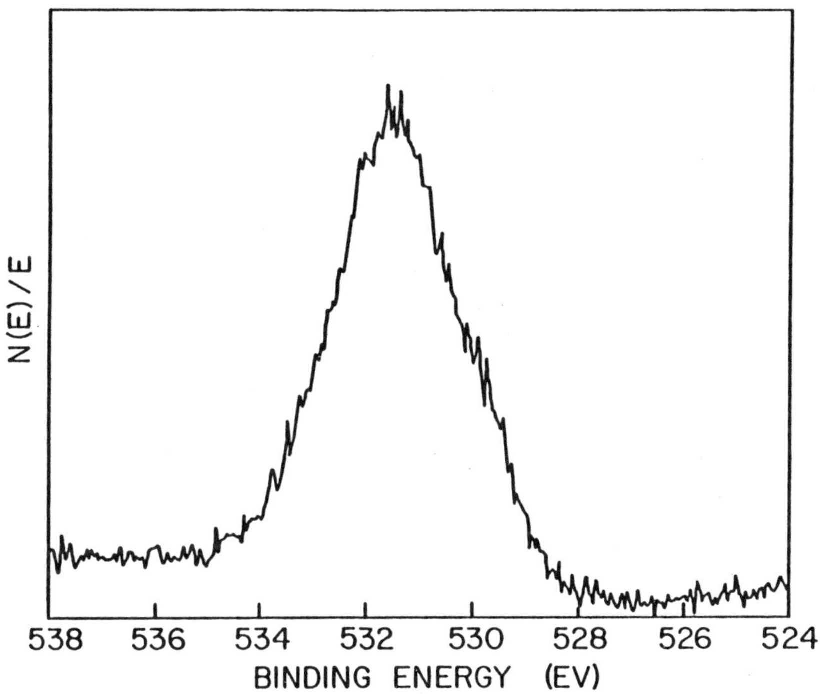

Figure III.4.3: High resolution XPS spectrum of a 752 Å film of PPVF on glassy carbon showing the O(1s) region. Pass energy = 25 eV.

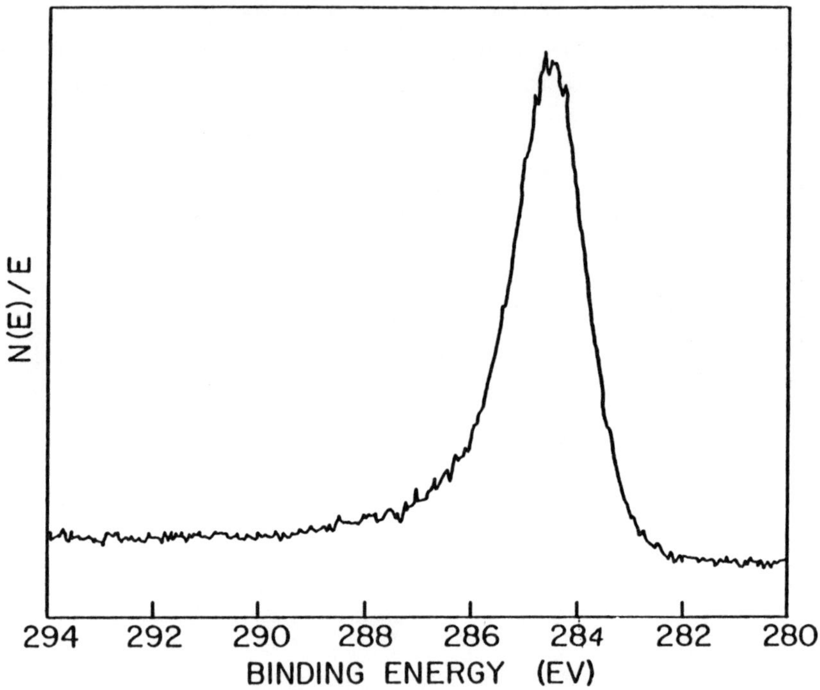

Figure III.4.4: High resolution XPS spectrum of a 752 Å film of PPVF on glassy carbon showing the C(1s) region. Pass energy = 25 eV.

pressure of oxygen in the reactor, it is not possible to determine if the oxygen is incorporated into the film during the discharge or if active sites in the film react with oxygen upon exposure to the atmosphere of if ferrocene in the PPVF film is oxidized by oxygen to yield ferricenium hydroxide (vide post). In the second case, a necessary requirement is that these active sites are relatively unreactive towards vinylferrocene or are inaccessable for reaction with vinylferrocene, since PPVF film covered substrates are post treated with vinylferrocene vapor for 0.5 h at 55°C prior to their removal from the plasma reaction chamber.

Examination of the high resolution XPS bands allows for some speculation as to the nature of the oxygen-containing functional groups present on the PPVF surface within the XPS analysis depth. The C(1s) spectrum shown in Figure 4 is asymmetric. The major portion of the carbon in the film is not bonded to electron withdrawing groups since the peak binding energy is that expected for C-H (284.5 eV). The asymmetry on the high binding energy side of the peak which tails out to ca 288 eV is suggestive of binding to electron withdrawing groups, these necessarily being oxygen since no other elements (besides Fe and H) are present in the PPVF film. The possible assignments for this range of chemical shift (284.5 - 288 eV) include C-OH, C-OR, and C=O which shift the C(1s) binding energy by 2, 2.5, and 3.5 eV respectively [3]. The existence of these oxygen containing functionalities is further evidenced by the O(1s) high resolution spectrum (Figure III.4.3).

Note that the peak is broad (FWHM = 2.5 eV) indicating more than one chemical state for oxygen. The peak is centered at 531.6 eV, which may be assigned to OH^- [3]. There is a pronounced shoulder at 530 eV which necessitates invoking the existence of O^{2-}. This binding energy is found for oxygen present in Fe_2O_3 [3]. This suggests that a small fraction of the ferrocene in the PPVF film, at least in the surface region, decomposed to Fe_2O_3. Alternatively, two ferricenium sites may share the oxygen dianion as a common counterion. On the high binding energy side there is also a shoulder at 533 eV, which can arise from OH, OR or C=O [3]. The high resolution Fe $2p_{3/2}$ spectrum shows a doublet, indicating two valences for iron, assigned to ferrocene and ferricenium as had previously been observed by us [1] and by others [2,4]. The presence of ferricenium requires a couterion, and this can account for the presence of OH^-. Deconvolution of the Fe $2p_{3/2}$ spectra gives 45% in the form of ferrocene and 55% in the form of ferricenium. Ross [2] had examined the relative "concentrations" of these as a function of depth by depth profiling PPVF films with intermitant high resolution XPS analysis. He observed a decrease in the relative intensity of the 711.6 eV band and an increase in the relative intensity of the 708.3 band as a function of depth. These results were interpreted to indicate that the ion beam damage processes were causing reduction of Fe(III) to Fe(II) (ferricenium to ferrocene). An equally valid conclusion would be that the Fe(III) exists only in the outermost layers of the surface.

A simple experiment to examine the relative concentrations of ferrocene/ferricenium immobilized on an electrode in the form of a polymer consists of measuring the film covered electrode potential with respect to a suitable reference electrode (potentiometry). Knowing the $E^{o'}$ of the couple and the otential - concentration relationship, the relative concentrations of the two forms of the couple are calculated in the entire polymer film, not only on the surface. Peerce and Bard [5] have previously demonstrated that redox polymer electrode potentials (PVF on Pt) are stable over time (measured potentiometrically) and the potential measured was related to the relative concentrations of the oxidized and reduced forms of the electroactive couple. The rest potential (the potentiometrically measured potential between the polymer film electrode and a suitable reference, e.g. Ag/AgCl) was measured for a series of PPVF films of various thicknesses after extraction and prior to other characterization by methods which might perturb the initial oxidation state of the PPVF film. This was exactly the same order of procedures used prior to XPS analysis. If the XPS determined relative concentrations of ferricenium and ferrocene reflected the bulk composition of the film, a potential anodic of the $E^{o'}$ of the PPVF film would have been observed (more anodic than 0.4 volts relative to Ag/AgCl (1.0 M KCl) in 0.1 M TBAP/CH_3CN). For all films examined, the rest potential was less than 0.17 volts (in 0.1 M TBAP/CH_3CN) irrespective of PPVF film thickness. Furthermore, the rest potential decreased as the PPVF film thickness increased. This is to be expected if the ferricenium in the PPVF film is confined to the outermost layers of the film whereby its contribution to the overall potential of the PPVF film would decrease as the film thickness increased. For the thinnest film examined (d(e) = 190 Å, rest potential = 0.17 V), the amount of ferricenium calculated to be presented in the film is less than 1.0% of the total amount of ferrocene present in the film. In the case of 749 Å film (rest potential = 0.08V), the amount of ferricenium initially present accounts for less than 0.2% of the total ferrocene centers present in the film. These rest potential measurements clearly show that the XPS determined ratio of ferricenium to ferrocene does not reflect this ratio in the bulk of the PPVF film.

The validity of the rest potential measurements used to determine the oxidation state of the film was demonstrated by potentiostatically charging a PPVF film (d(e) = 800 Å) at 0.45 V so that the film was 55% in the ferricenium state ($Q_{injected} = 4.6 \times 10^{-4}$ C, $Q_T = 8.4 \times 10^{-4}$ C). The auxilliary electrode was then disconnected and the charged film removed from the cell. The film was then air dried and stored for 5 days. After 5 days, the rest potential was measured and was 0.4 volts. The charged film was then potentiostatically poised at 0.0 V, whereby 75% of the initial charge injected was recovered (determined coulometrically). Less than 100% charge recovery was probably due to partial discharge of the film due to the finite time required to remove the film from solution. At bare glassy carbon electrodes, the onset of the irreversible background redox reaction observed with no polarizers in solution (0.1 M TBAP/CH_3CN) was slightly anodic of 0.4 V. Presumably, the film might have partially discharged by oxidizing the contaminent in solution which gives rise to the background currents observed. Rest potentials of films that were charged to give potentials between 0.4 to 0.0 volts were stable (i.e. did not drift) for the measurement times examined (~10 minutes). Films whose redox state afforded rest potentials more anodic than 0.4 volts. In spite of the partial discharge of the PPVF film above, the charged film rest potential experiment clearly indicates that PPVF films can hold charge and that the fractional amount of the total film charged is reflected by its rest potential.

It is instructive to calculate the thickness of the partially oxidized layer on the PPVF film. Using the rest potential determined fractional amount of the film that is oxidized, and assuming that the oxidized layer is a uniform layer of ferricenium on top of a layer of polymer which is 100% ferrocene (which is unlikely to occur), then an oxidized layer of ca 2 Å is calculated for the case of the 190 Å film and 1.5 Å is calculated for the case of the 750 Å film. These thicknesses suggest that only the top "equivalent monolayer" of the film was oxidized.

The above results imply that the films are not oxidized during plasma deposition, rather the surface was oxidized as a consequence of exposure to the atmosphere. Umana _et al._ [4] have examined ferrocene modified electrodes by XPS prepared by various methods, including plasma polymerization of vinylferrocene. They reported that the various modification schemes used affected the relative amounts of ferrocene to ferricenium present on the ferrocene modified electrodes. In general, ferricenium was present on the modified surfaces as evidenced by XPS, and reasons for ferricenium formation attributable to each modification scheme were proposed. (8 different modification schemes were used, including covalent bonding and adsoprtion of ferrocene polymers. One of the modification schemes was plasma polymerization of vinylferrocene and several variations of deposition reaction parameters were examined.) Unfortunately, Umana _et al._ leave the readers with the impression that the coupling schemes used affect the oxidation state of the entire modified layer, but the results presented in this dissertation show that generalization of bulk composition of polymer modified electrodes cannot be made using XPS as the only analytical technique. In a later paper, Umana _et al._ [6] used Grazing Angle Esca to determine film thickness of PPVF and PVF modified electrodes. Had this technique been used to examine the ferrocene/ferricenium ratio reported in the earlier paper, more definitive statements of the distribution of these ratios as a function of depth could have been made.

Ross [2] had previously proposed that the ferricenium found on PPVF surfaces was a result of air oxidation of ferrocene according to:

(1) $2FeCp_2 + 1/2\ O_2 + H_2O \longrightarrow 2(FeCp_2^+ - - - OH^-)$

This hypotehsis is supported by the observation in this thesis that Fe(III)/Fe(II) ratios and O/C ratios (determined by XPS) were higher on PPVF films analyzed 1 month after their preparation (compared to PPVF films analyzed within a day after their preparation). It was also observed that the rest potential of PPVF film electrodes measured one month after their preparation were more anodic when compared to PPVF film electrodes prepared in the same deposition examined within one day after preparation.

Although the ferricenium is present only on the outermost surface, the amount of ferricenium present on the film cannot account for all the oxygen present, since the oxygen is present in concentrations 5 to 10 times greater than the ferricenium present. This oxygen is most probably bound to carbon in the film. The XPS results cannot determine whether oxygen is present throughout the film. Depth profiling experiments discussed elsewhere in this report suggest that although the relative concentration of oxygen is reduced, it is present through the bulk of the film. However, these depth profiling results are inconclusive as to the chemical state of the oxygen and the relative concentrations of oxygen as a function of depth due to possible preferential sputtering of these oxygen moieties [21].

Bulk Characterization

Fourier transform transmission infrared spectroscopy was used to evaluate the bulk chemical composition of PPVF films. For this purpose, relatively thick films of PPVF deposited on silicon substrates were required in order to attain a sufficient infrared signal intensity. Figure III.4.5 shows the infrared spectrum of an as deposited PPVF film on silicon (d(u) = 5360 ±60Å), Figure III.4.6 is the spectra of the same film after extraction (d(e) = 4200 ±80Å) and for comparative purposes, Figure III.4.7 shows the infrared spectrum of authentic polyvinylferrocene (on KBr).

Qualitatively, all of the spectra show similar features. Bands expected for ferrocene include[7]: C-H stretch (aromatic) at $\bar{\nu}$ = 3085 cm^{-1}, antisymmetric C-C stretch at $\bar{\nu}$ = 1411 cm^{-1}, antisymmetric ring breadth at $\bar{\nu}$ = 1108 cm^{-1}, C-H bend (||) parallel to the plane of the cyclopentadienyl ring) at $\bar{\nu}$ = 1002 cm^{-1}, C-H bend ($\perp$) (perpendicular to the plane of the cyclopentadienyl ring) at $\bar{\nu}$ = 811 cm^{-1}. Each of these bands is observed in the spectra of PVF and unextracted and extracted PPVF. These spectra unequivocally demonstrate that PPVF is a ferrocene containing polymer. There are some differences, however, between the PPVF spectra (extracted and unextracted) and the PVF spectra. In the PPVF spectra, the presence of a band in the 3600 cm^{-1} frequency range suggests -OH groups are present in the film. Since these spectra were taken in vacuo, contribution to the

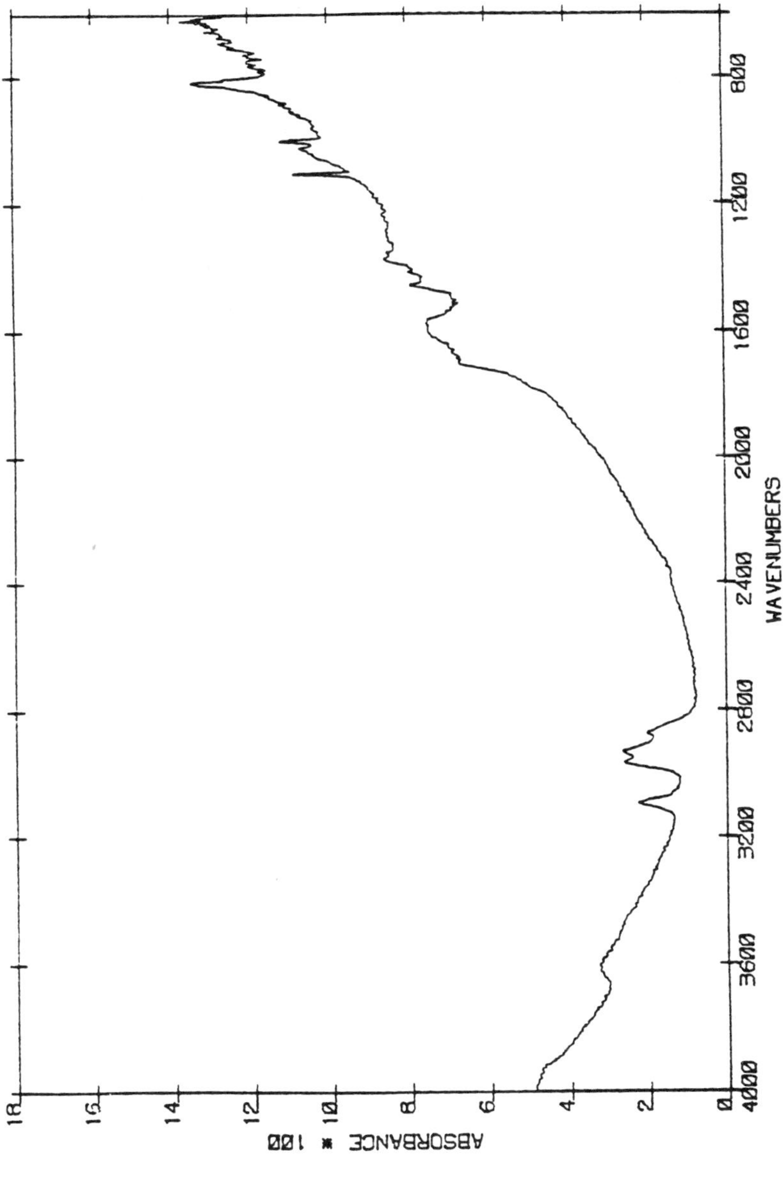

Figure III.4.5: IR spectrum of PPVF deposited on silicon. Unextracted film thickness = 5360 ±60 Å. Curvature in base line is due to difference in reflectivity between PPVF film covered silicon and bare silicon.

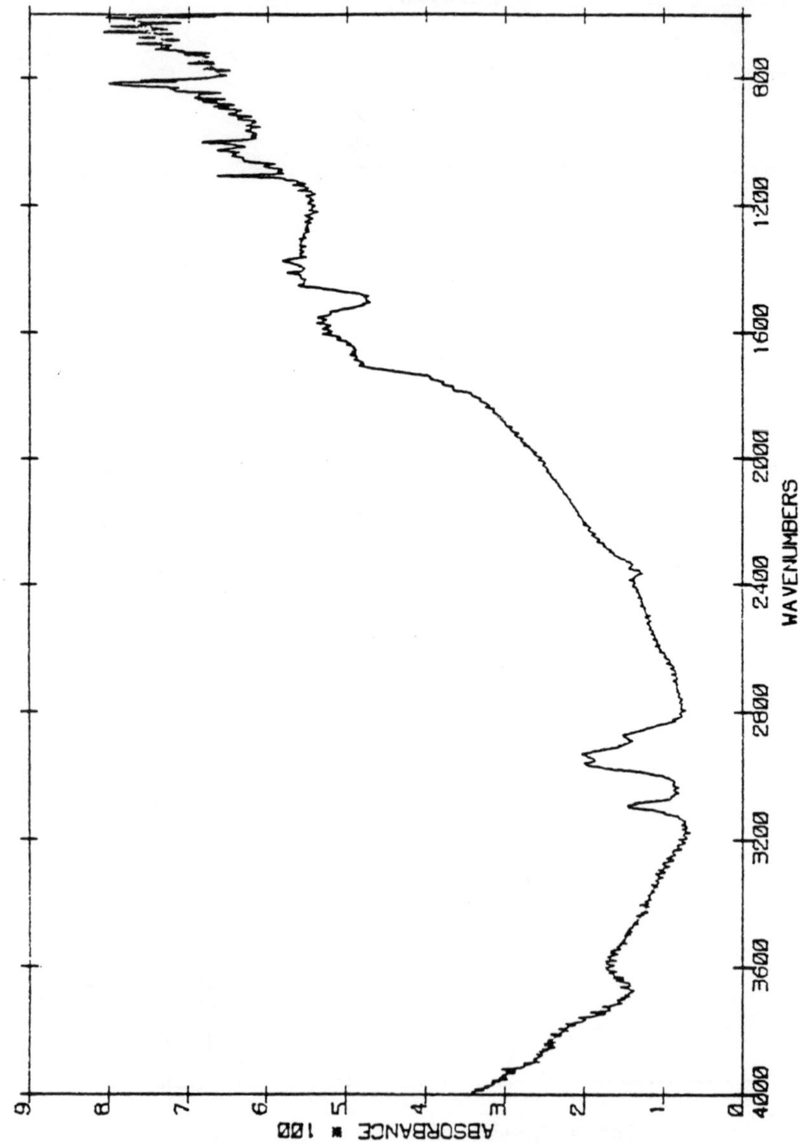

Figure III.4.6: IR spectrum of PPVF deposited on silicon. Same film as shown in Figure 5, after extraction. Extracted film thickness = 4200 ±80 Å.

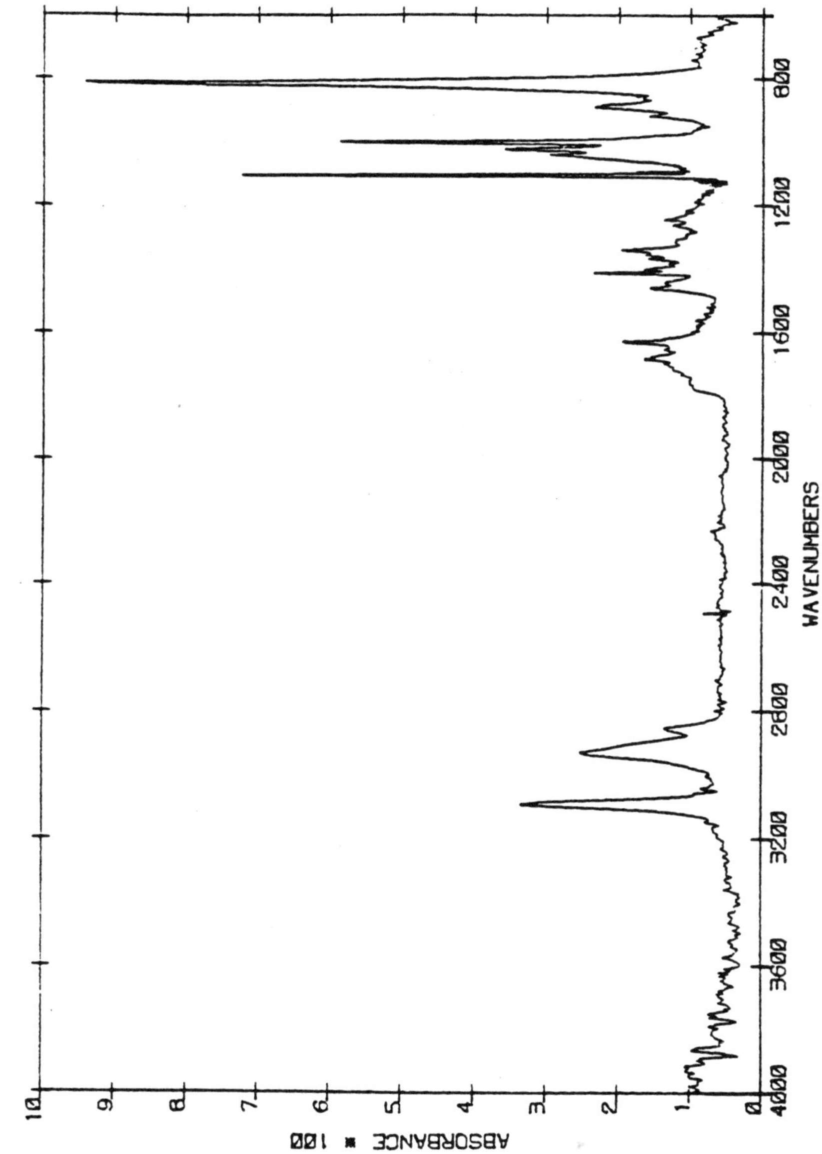

Figure III.4.7:IR spectrum of polyvinylferrocene adsorbed on KBr.

absorption at this frequency by adsorbed water should be minimal. The absence of this band in the PVF spectra supports this conclusion. Examination of a standard spectrum [8] of α-hydroxy ethyl ferrocene (structure shown below)

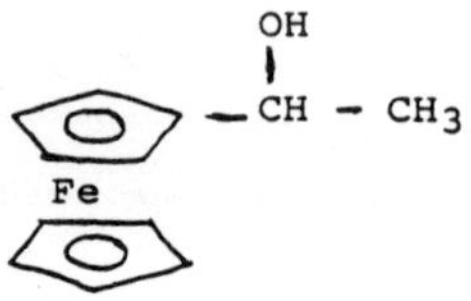

(I)

a free O-H stretching band and a broad band for hydrogen bonded O-H stretching was shown at 3650 cm^{-1} and 3400 cm^{-1} respectively. Therefore, if the assignment of O-H present in PPVF films is correct, the O-H groups (possibly bonded a to the ferrocene cyclopentadienyl ring) exist as free O-H groups. Although very crude, the fractional amount of α-hydroxy substituted ferrocene groups present in the film may be estimated by comparison of the intensity ratio of the OH- absorption band to other bands in the spectra between the model compound (I) and the PPVR films. The fractional amount of α-hydroxy substituted ferrocenes was estimated to be 30-40% of the ferrocene present in the PPVF film (unextracted and extracted). This represents a lower limit, since the O-H band in the model compound was much broader than other bands present and the peak heights of the bands were used for the calculation. Nevertheless, this crude estimation suggests that PPVF films contain a significant amount of hydroxy-substituted ferrocenes. This finding supports depth profiling experiments of PPVF films where the O/Fe ratio was found to be 0.8 in the bulk of the PPVF film.

In the 1800-1600 cm^{-1} region for the PPVF spectra, a prominent, but poorly resolved absorption band was present. The intensity ratio of this band relative to ferrocene bands in the PPVF spectra is ca 10 times greater than this ratio for PVF spectra. Carbonyl groups α to ferrocene absorb in this region [7]. However, vinyl groups absorb in this region also. That part of the absorption in this region might be a consequence of α-carbonyl grougs in PPVF is consistant with the cyclic voltammetric responses of PPVF film electrodes. Figure III.4.8 shows the cyclic voltammogram of a 752 Å PPVF film on glassy carbon in 0.1 M TBAP/acetonitrile. A reduction peak is observed at ca 0.62 volts vs. Ag/AgCl, which is shifted ca 200 mV of the $E^{o'}_{app}$ of the main redox couple. This observation suggests two redox couples are present in the PPVF film. Substituents on ferrocene can shift the $E^{o'}$ significantly. However, the shift in $E^{o'}$ for α-hydroxy and vinyl groups is very small (ca 20 mV) [9-12]. α-carbonyl groups shift the ferrocene $E^{o'}$ by ca 200 mV anodic of $E^{o'}$ of unsubstituted ferrocene. The magnitude of this shift compares favorably with the potential difference between the two waves in the cyclic voltammogram in Figure 8. Note, however, that the anodic wave is not observed for the higher potential couple. This may be accounted by for the broadness of the main wave (FWHM = 140 mV). Also, the higher potential oxidation can be

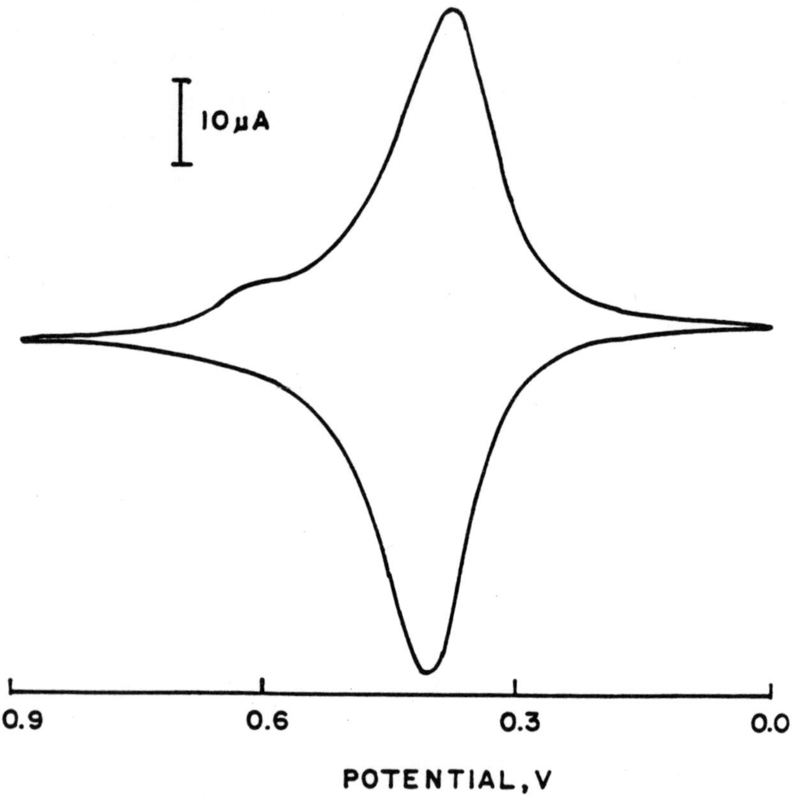

Figure III.4.8: Cyclic voltammogram of a 750 Å PPVF film on glassy carbon in 0.1 M TBAP/CH_3CN. Sweep rate = 10 mV-s^{-1}, electrode area = 0.24 cm^2.

efficiently mediated by the lower potential couple [13], especially since its relative concentration is much greater, whereby a distinct anodic wave is not observed for the higher potential couple. The higher potential couple was observed for films of all thicknesses and in all solvent-supporting electrolytes examined, and the magnitude of the peak current relative to the main peak current was comparable in all cases. Based on both the cyclic voltammetry and the infrared data, the assignment of α-carbonyls present in the film is made with some confidence. Based on the cyclic voltammetry, an estimate of ca 10% of the ferrocene centers in the PPVF film contain a-carbonyl groups.

The cyclic voltammetric data does not exclude the possibility that β-carbonyl groups or C=C groups are present in the film. β-carbonyl groups would not significantly shift the $E^{o'}$ of ferrocene. C=C groups have been observed via infrared spectroscopy in plasma polymerized films derived from unsaturated hydrocarbons [14]. Incorporation of carbonyl groups into plasma polymerized films upon exposure to oxygen has been generally observed for films derived from various monomers [15].

Further insight into the structure of PPVF films is afforded by close examination of the "structurally significant" bands of ferrocene. These are the bands at 1108 cm^{-1} and 1002 cm^{-1}. The absence of these bands in spectra of ferrocene derivatives indicate that both cyclopentadienyl rings are substituted [7]. The presence of these bands in the case of a ferrocene derivative indicates that only one of the ferrocene cyclopentadienyl groups is substituted. The spectral region between 1200 cm^{-1} and 750 cm^{-1} is shown in Figure III.4.9 for unextracted PPVF and PVF. Note that the 1108 cm^{-1} and 1002 cm^{-1} bands are present in both PPVF and PVF. Furthermore, the intensity ratios of the 1108 and 1002 cm^{-1} bands to the 811 cm^{-1} band are 0.77 and 0.59 respectively for the PPVF film, which compares very favorably with these ratios for PVF (0.78 and 0.60). The 1002 cm^{-1} to 811 cm^{-1} ratio for PPVF and PVF (this work) are in good agreement with the ratio of these band intensities (0.61) reported by Aso et al. for PVF synthesized by either free radical or cationic polymerization of vinylferrocene [16]. This strongly suggests that the cyclopentadienyl ring which is originally unsubstituted in vinylferrocene remains unsubstituted in the PPVF films. Figure III.4.10 compares the PPVF film before and after extraction whereby the spectral features are shown to be almost identical in both cases, being of lower intensity in the latter. The 1108 cm^{-1} and 1002 cm^{-1} to 811 cm^{-1} band ratios are 0.73 and 0.61, respectively, in good agreement with the unextracted PPVF and the PVF ratios. This shows that the aforementioned correspondence between the unextracted PPVF film and PVF was not a consequence of the unextracted material in the film. This also suggests that the material which is extracted from the film is very similar to the material which remains unextracted, except that it is probably a lower molecular weight oligomer. Ferrocene derivatives which have

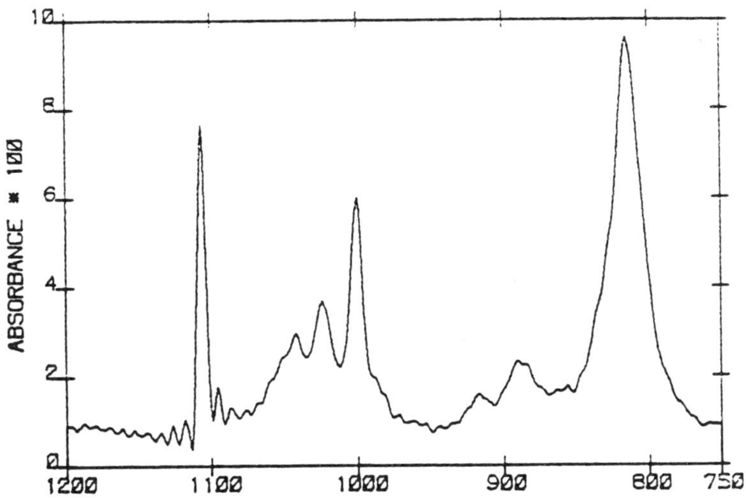

Figure III.4.9A: IR spectrum of PVF on KBr from 1200 - 750 cm^{-1}.

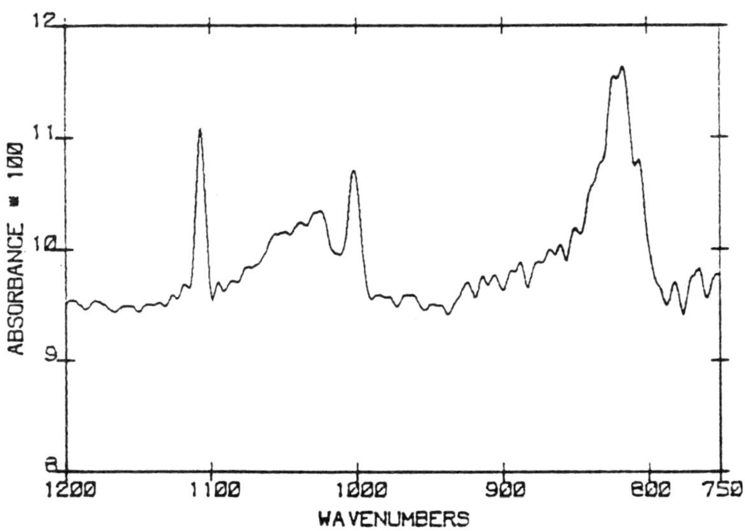

Figure III.4.9B: IR spectrum of unextracted PPVF film on Figure III.4.6 from 1200 - 750 cm^{-1}.

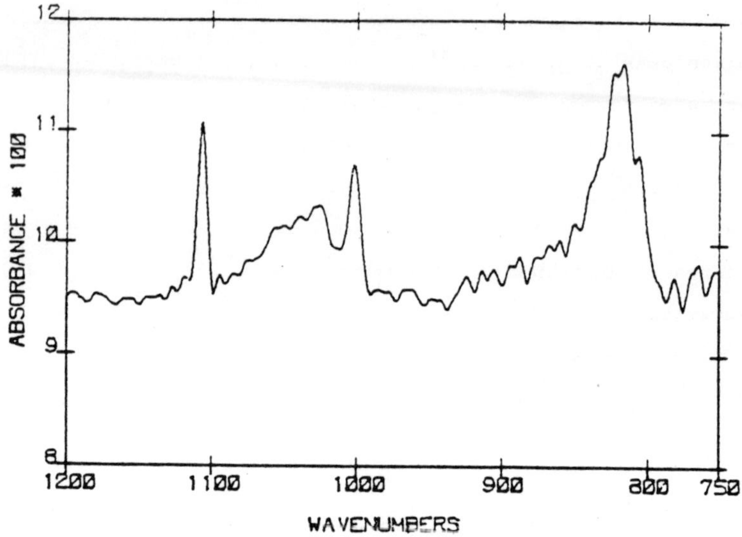

Figure III.4.10A: IR spectrum of unextracted PPVF. Same spectrum as Figure III.4.9B.

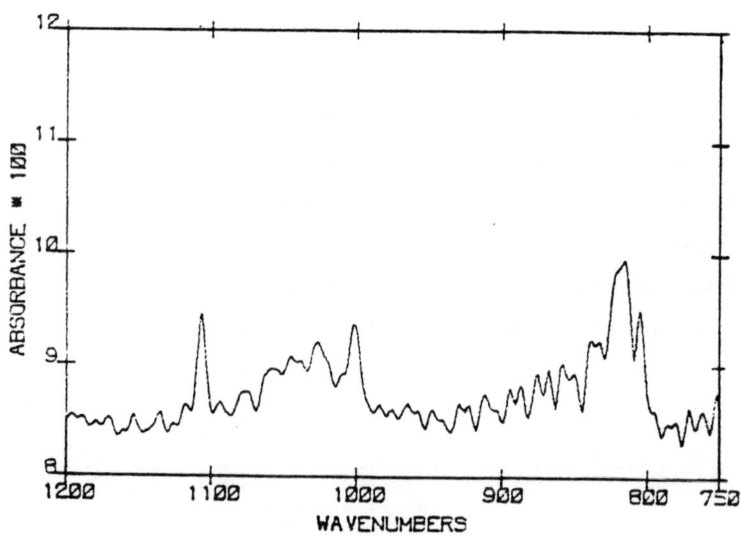

Figure III.4.10B: IR spectrum of PPVF film in III.4.10A after extraction.

two subsitutents on one ring generally show a singlet or a doublet absorption peak at ca 920 cm^{-1} for 1,2 and 1,3 homoannular disubstituted ferrocenes, respectively [7]. Within the limits of the signal to noise ratio, these bands are not observed in the PVF spectrum or either of the PPVF spectra. The implication of these observations in the 1200-750 cm^{-1} spectral region are two-fold: (1) The resultant PPVF films are mainly composed of mono-substituted ferrocene polymer, most probably with ferrocenes bonded to each other through an alkyl chain; (2) the crosslinking of the PPVF film predominantly occurs through the alkyl side chain, with little or no contribution from crosslinking via direct bonds to cyclopentadienyl rings on ferrocene.

The ratios of the intensities 1108, 1002 and 811 cm^{-1} bands of the extracted PPVF film to the intensities of these respective bands in the unextracted film give an average value of 0.81. This ratio, which is the fraction of the film left after extraction is in good agreement with that calculated by ellipsometry (0.78).

In the C-H stretching region (3200-2700 cm^{-1}), some very important differences between PVF and PPVF are noted. Figure III.4.11 shows this spectral region for the unextracted PPVF film and PVF. In PVF, the predominant C-H stretch is that for C-H on the cyclopentadienyl ring. The bands at 2930 and 2855 are due to $-CH_2$ stretching [8]. In the PVF spectrum, there is very little absorption due to $-CH_3$ stetching, as would be expected for linear PVF with a high degree of polymerization, since $-CH_3$ groups

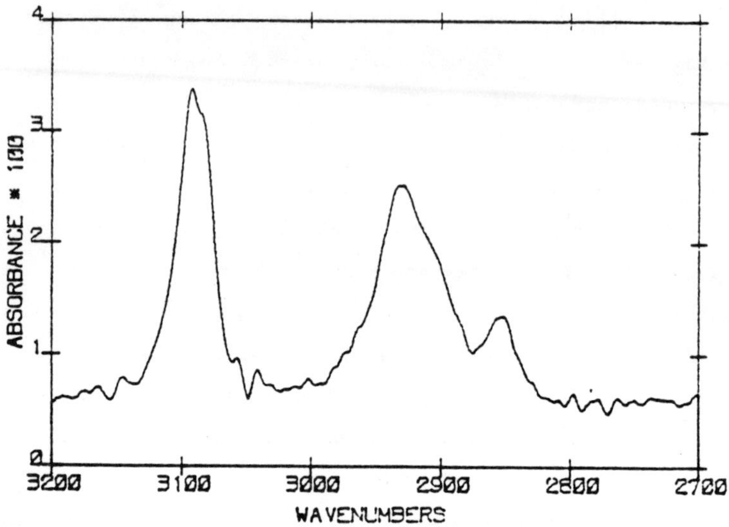

Figure III.4.A: IR spectrum of PVF in the C-H stretch region.

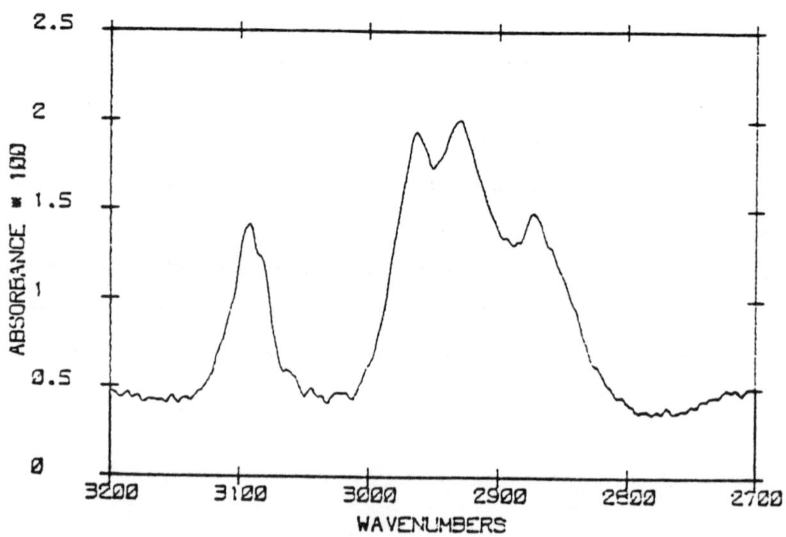

Figure III.4.B: IR spectrum of unextracted PPVF in C-H stretch region.

present would be terminal $-CH_3$ groups. The relative number of these groups increases as the degree of polymerization decreases. However, $-CH_3$ bands are very prominent in the PPVF spectra (2965 and 2875 cm^{-1})[17]. Also, the $-CH_2$ bands are more intense relative to the cyclopentadienyl C-H band. These large differences in the C-H stretch region between PPVF* and PVF suggest the alkyl side chains are predominant participants in the plasma polymerization of vinylferrocene. If free radicals are the major precursors in the plasma polymerization of vinylferrocene, and the mechanism of polymerization is free radical addition (as for conventional polymerization *via* free radicals), the observation of this relatively large fraction of methyl groups present in the PPVF film would suggest a lower molecular weight linear polymer had formed (i.e. termination by disproportionation only). However, this is not the case, since the extracted PPVF film spectrum is identical to the unextracted PPVF film spectrum (Figure III.4.12). Extracted PPVF films were insoluble in all solvents examined (*vide post*). This implies the films are highly crosslinked. One way to rationalize the abundance of $-CH_3$ groups and crosslinking is that hydrogen abstraction by free radicals in group polymer chains from other polymer chains and subsequent chain addition is very important (i.e., chain transfer). Alternatively, as proposed by Dilks and Kay [18], carbene (e.g. methylene) may be an important precursor or participant in the polymerization process. The methylene fragment can insert into

*Incorporation of methyl groups in plasma polymerized films is not unique to PPVF and has been found *via* infrared spectroscopy for plasma polymerized styrene [3].

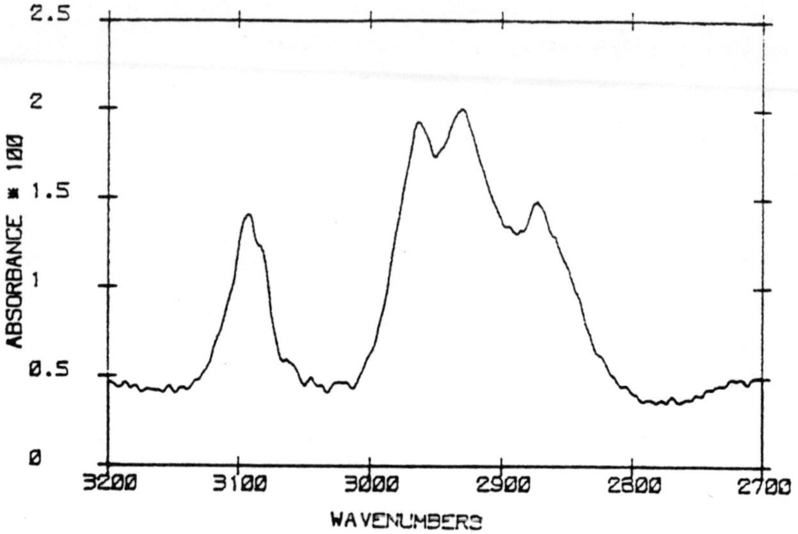

Figure III.4.12A: Same IR spectrum as Figure III.4.11B.

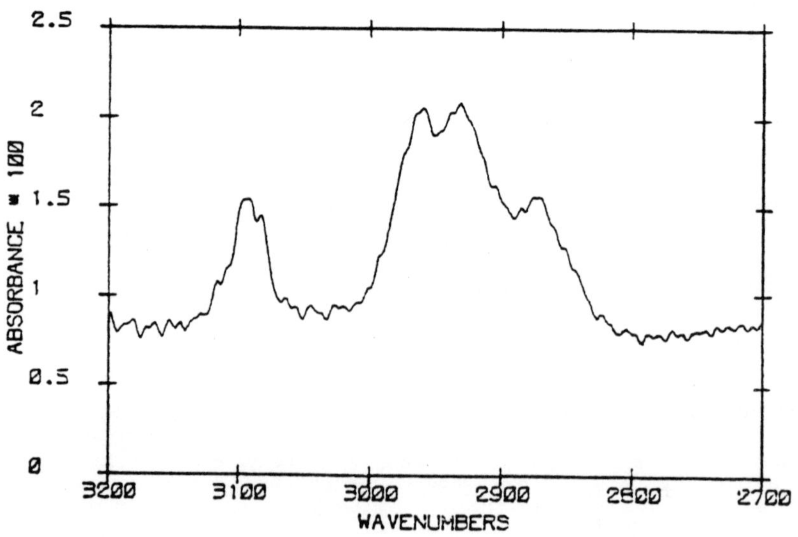

Figure III.4.12B: IR spectrum of PPVF film in Figure III.4.12A after extraction.

C-H bonds to give $-CH_3\cdot$.* It may also play a role in crosslinking of polymer chains, either by hydrogen abstraction or by addition between two polymer chains.

At present, speculation as to which crosslinking mechanism predominates is not possible. In either case, however, hydrogen abstraction is implicated as an important intermediate step to crosslinking in the polymer.

Solubility and Stability of PPVF Films

The solubility of extracted PPVF films was examined in various solvents. Prior to this examination, a 750 Å PPVF film (extracted) was extracted again in acetonitrile (8 hrs) and the film thickness measured ellipsometrically. No difference in film thickness or refractive index was observed after the second extraction. This film was then soaked in methylene chloride for 12 hours. The color of the electrode was identical to the color observed before soaking in methylene chloride (metallic blue). The lack of any observable color change indicates the film had not diminished in thickness and was not soluble in methylene chloride. PVF of molecular weights greater than ca 75,000 g/mole is very soluble in methylene chloride. The above experiment demonstrates PPVF is highly crosslinked. The electrode was then ultrasonically agitated in methylene chloride for 10 minutes. Again no color change was observed. The 10 minute ultrasonic procedure was then used for the following solvents (in sequence) with the same electrode as above: benzene, acetone, ethanol, chloroform, carbon tetrachloride, water and methanol. No color change was observed at any point and the PPVF film was not soluble in these solvents. Had the PPVF film decreased in thickness, the following color changes would have occured in order to decreasing thickness: metallic blue, deep royal blue, purple, brown, tan, and finally, black, which is the color of the glassy carbon electrode. These results are in contrast to that observed by Murray with PPVF films formed with Argon as a diluent [19]. He reports that PPVF films are quite soluble in methylene chloride.

In the course of the solubility studies above, it was observed that nonpolar solvents did not wet the PPVF surface while polar compounds wet the film surface. This implies that the surface is polar, which is to be expected by the ionic character shown to be on the surface from the XPS results and the carbonyl and hydroxyl groups fround by infrared spectroscopy.

To test the adherence of PPVF films to glassy carbon electrodes, scotch tape was firmly pressed against the PPVF film and pulled off. This was repeated several times, but the film remained intact. The only method found so far to remove PPVF films from electrodes was to decompose the film using organic or inorganic oxidants in concentrated sulfuric acid.

Extracted PPVF films of varying thickness were evaluated for possible swelling of the film by acetonitrile. The films were placed in a beaker of acetonitrile and examined for color change. No discernable change was observed for films less than 1000 Å, a slight color change was observed for a 1400 Å (light yellow green to gold) and a pronounced color change was observed for a 4560 Å film (beige to deep red). In the latter case, after removing the film from the acetonitrile, the color reverted to its original color. Hence, some swelling of the PPVF film occurs, although the extent of swelling is not known. Based on the results for PPVF films less than 1000 Å thick, the swelling is probably less than 10% of the film thickness, otherwise a color change would have been observed.

The electrochemical stability of the films markedly depended on the supporting electrolyte used. PPVF films were very stable in TBAP/acetonitrile, TEApTs/acetonitrile, TBAP/methylene chloride, unbuffered neutral $LiClO_4$/water, and unbuffered neutral KNO_3/water. Less than 8% of the electrochemically active sites in the PPVF films examined with the above solvent-supporting electrolytes after extensive electrochemical cycling (electrochemical experiments were generally conducted for periods in excess of 3 h).

PPVF film electrodes were considerably less stable electrochemically in $TBAP_6$/acetonitrile, unbuffered neutral KF/water, and unbuffered neutral KCl/water. The reasons for this instability are not clear; apparently the counterion plays a dominant role in determining the stability of ferricenium in PPVF films.

With one exception, PPVF films of the thickness studied were reflective after electrochemical characterization in SSE in which the films were stable. The amount of light scattered from the electrolyzed portion of the film was indistingishable from the unelectrolyzed area. This allowed film thickness measurement after electrochemical characterization. Except for methylene chloride, no significant change in film refractive index or thickness of the film was measured. In the case of methylene chloride, a 750 Å film decreased in thickness by 20 Å (2.6% change). Thus in general, the observed loss in electrochemical activity was not due to significant desorption of the polymer or oligimers therein.

Electrochemical cycling of the thickest film studied (d(e) = 4560 ±60 Å) caused the electrolyzed portion of the film to extensively scatter laser light. The electrolyzed area of the film was easily visible with the naked eye and various colors were observed in this region. However, the unelectrolyzed portion of the film remained optically smooth.

The above film was examined by electron microscopy, along with thinner films (1400 Å, 750 Å, 500 Å) after electrochemical cycling. In the case of 1400 Å or less thick films, no features could be seen, and the electrolyzed area was indistinguishable from the unelectrolyzed area. These PPVF film covered surfaces (both glassy carbon and silicon) were so smooth and featureless that the same problem in finding a feature on which to focus the microscope was encountered as in the case of bare glassy carbon surfaces (see Experimental Section). In contrast, the thickest film examined showed a great deal of damage occured as a consequence of electrochemical cycling. The SEM of this film is shown in Figure III.4.13. The unelectrolyzed area shows no features at the highest usable magnification. The electrolyzed area however shows that the polymer film as literally torn and fractured as a consequence of electrochemical cycling. This is presumed to have occurred because of the electrochemically induced swelling, causing stress, strain and eventual rupture in the film as well as loss of adhesion to the glassy carbon substrate. That this did not occur in thinner films may be due to the absolute change in thickness not being sufficient to cause appreciable stress and strain in the film. Nevertheless, these SEM results give strong support for the electrochemically induced swelling of the films.

Electron microprobe analysis of regions where the polymer film appeared to be completely sheared away from the electrode surface showed very little iron present in these regions whereas the same analysis over the film showed nearly a 50 fold increase in the iron signal intensity. This suggests that the film was literally peeled off the glassy carbon surface. Estimation of the PPVF film thickness by measuring a cross section of the sheared polymer layer afforded a film thickness of 4400 Å, which considering the accuracy of such a measurement, is in excellent

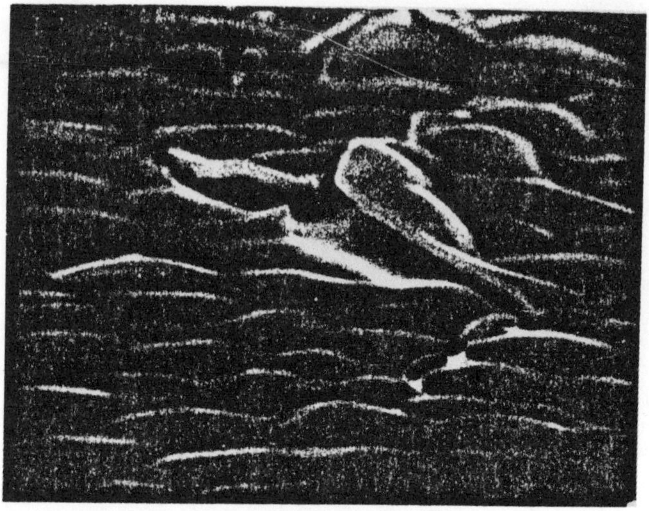

A: Electrolyzed area

4200 X

B: Electrolyzed area

875 X

Figure III.4.13: Scanning Electron Micrographs of 4560 ±60 Å PPVF film on glassy carbon (extracted films). Beam voltage = 5 KV, tilt angle = 65°.

C: Electrolyzed area

4200 X

D: Unelectrolyzed area

17,000 X

Figure III.4.13: Scanning Electron Micrographs of 4560 ±60 Å PPVF film on glassy carbon (extracted films). Beam voltage = 5 KV, tilt angle = 65°.

agreement with the ellipsometrically determined film thickness (4560 Å).

Electrochemical characterization of PPVF films on glassy carbon after XPS characterization (in which the PPVF surface was illuminated with Mg K_{α} x-rays in excess of 2 h) showed no x-ray induced damage or modification of the PPVF film. The coulometrically and ellipsometrically determined concentration of the ferrocene present in the films was identical to that found for a film made in the same deposition run and electrochemically characterized without prior XPS characterization. Furthermore, the charge transport kinetics (as evidenced by the peak separation as a function of sweep rate) were equivalent for both films. Therefore, for the XPS analysis time studied here, PPVF films are stable to soft x-ray bombardment and photoelectrons which are generated within the film.

Conclusions

PPVF films which have uniform thickness and reproducible optical properties have been deposited in a neat vinylferrocene plasma. Steady state plasma pressure was achieved by control of RF power, with other parameters held constant.

It has been shown that it is crucial not only to examine the properties of the "as deposited" film, but more important to examine the film properties after extraction. These "extracted film" properties strongly depend on steady state plasma pressure.

A greater fraction of the "as deposited" PPVF film is an insoluble polymer at the lowest steady state plasma pressure studied.

Using ellipsometric film thickness measurements and coulometric coverage determination, it was established that films which were of different thickness (different depositions) had equivalent densities as reflected in the calculated concentration of ferrocene centers in the film. These equal density films were made using steady state plasma pressure of 6-8 mtorr. The deposition rate and the properties of the deposited film are found to be independent of the substrates examined.

The surface composition of PPVF films in terms of fraction of the ferrocene which is oxidized is different from the bulk of the film. Comparison of XPS results to electrochemical rest potential measurements of PPVF films indicates that ferricenium is present only in the outermost layers of the film, probably confined to not more than the outermost 1 or 2 equivalent monolayers.

Infrared spectroscopy and cyclic voltammetry show that considerable oxygen is present throughout the film. These results strongly suggest that the oxygen is predominantly in the form of hydroxyl and carbonyl functional groups. Infrared spectra unequivocally show that PPVF films are ferrocene polymers. The structurally important bands strongly suggest that the cyclopentadienyl rings are not substituted during polymerization and hence play a minor role in the plasma polymerization process. The high fraction of $-CH_2$ and $-CH_3$ groups present in PPVF (compared to PVF) suggest considerable rearrangement and fracture of the alkenyl sidechains of vinylferrocene occur during plasma polymerization. The importance of carbene in the polymerization and crosslinking of PPVF is strongly implicated. Extracted PPVF films are insoluble in all solvents examined. They are very stable and adhere strongly to the surface. Swelling of the film by acetonitrile does occur as shown by color changes of PPVF in acetonitrile and as suggested by SEM of the thickest film. In the latter experiment rupture of the film occured presumably as a consequence of electrochemically induced swelling. For comparison, Schroeder *et al.* [20] have observed *via* SEM gross changes of a similar sort in polymer morphology as a consequence of electrochemical cycling of thick films (2500 Å) of tetrathiofulvalene polymers. They argue that the observed changes in morphology are a consequence of activated diffusion of counterions through the film, changing the free volume which in turn alters the polymer morphology. That counterions diffuse through the film with difficulty is supported by kinetic properties of these films and transport of electroactive couples through the film.

References

1. Dautartas, M. F. and Evans, J. F., J. Electroanal. Chem., 109, 301 (1980).
2. Ross, M. R., Ph.D. Dissertation, March 1981, University of Minnesota, Minneapolis, MN.
3. Wagner, C. D., Riggs, W. M., Davis, L. E., Moulder, J. F. and Muelenberg, G. E., Handbook of X-Ray Photoelectron Spectroscopy, Perkin-Elmer Corp., Physical Electronics Division, Eden Prairie, MN, 1979.
4. Umana, M., Rolison, D. R., Nowak, R., Duam, P. and Murray, R. W., Surface Science, 101, 295 (1980).
5. Pierce, P. J. and Bard, A., J. Electroanal. Chem., 108, 121 (1980).
6. Umana, M., Denisevich, P., Rosiloen, D. R., Nakahama, S. and Murray, R. W., Anal. Chem., 53, 1170 (1981).
7. Rosenblum, M., Chemistry of the Iron Group Metallocenes, Part I, John Wiley & Sons, New York, 1965.
8. Sadtler Research Laboratories, Inc., Spectra 44875 P.
9. Page, J. A. and Willkinson, G., J. Amer. Chem. Soc., 74, 6149 (1952).
10. Hoh, G.L.K., McEwen, W. E. and Kleinberg, J., J. Amer. Chem. Soc., 83, 3949 (1961).
11. Hennig, H. and Gurtler, O., J. Organometal. Chem., 11, 307 (1968).
12. Gorton, J. E., Lentzner, H. L. and Watts, W. E., Tetrahedron, 27, 4353 (1971).

13. Anson, F. C., J. Phys. Chem., 84, 3336 (1980).

14. Shen, M. and Bell, A. T., "A Review of Recent Advances in Plasma Polymerization", pp. 1-33 of Ref. 44.

15. Hollahan, J. R. and Bell, A. T., Techniques and Applications of Plasma Chemistry, Wiley, New York, 1974.

16. Aso, C., Kunitake, T. and Nakashima, T., Die Makromoledulare Chemi, 124, 232 (1969).

17. Colthup, W. B., Daly, L. H. and Wiberley, S. E., Infrared and Raman Spectroscopy, Academic Press, Inc., New York, NY, 1975.

18. Dilks, A. and Kay, E., Macromol., 14, 855 (1981).

19. Nowak, R. J., Schultz, F., Umana, M. and Murray, R. W., Anal. Chem. 52, 315 (1980).

20. Schroeder, A. and Kaufman, F., J. Electroanal. Chem., 113, 209 (1980).

21. Ullevig, D. M. Evans, J. F., Anal. Chem., 52, 1467 (1980).

III.5 X-Ray Photoelectron Spectroscopic and Secondary Ion Mass Spectrometric Thin Film Analysis of PPVF

Sputtering Behavior of PPVF

The theoretical and practical details of the use of the QCM in sputtering studies of organic films have been given by Ullevig and Evans[1] and references therein. Only a brief description of the technique will be presented here. The microbalance consists of a quartz crystal, each face of which is coated with a relatively inert metal (Ag or Au). This quartz resonator is driven by a Pierce oscillator circuit. As material is added to (or removed from) one of the faces, the frequency of resonant oscillation of the composite will change as if an equivalent mass of quartz were hypothetically inserted between (or removed from) the electrodes. This linear relationship was experimentally determined by Sauerbrey[5] and found to hold up to a 10% loading limit. The relationship between mass loss or gain from a face of the QCM (Δm) and the resulting frequency change (Δf) are formalized in the Sauerbrey equation:

$$\Delta f = \frac{f^2}{\rho NA} \Delta m \qquad (1)$$

where: Δm is the mass change (g), A is the active area of the crystal (cm^2), ρ is the density of quartz (2.65 g cm^{-3}), Δf is the frequency change associated with Δm (Hz), f is the resonant frequency of the crystal (Hz), and N is the frequency constant of quartz (1.67 x 10^5 Hz-cm). While it appears that the terms on the right hand side of eqn (1) with the exception of Δm can be used to define a constant relating frequency change and mass change, as Ullevig has ponted out, this cannot be done when mass is not uniformly removed from all differential elements of the active area.[(1)] Since the experimental arrangement used in the present study was exactly the same as that used by Ullevig, the same correction factor determined by numerical analysis was used here. This factor accounts for both the nonuniform radial sensitivity of the QCM and the radial nonuniformity in the ion beam flux (see experimental).

Of fundamental interest in sputtering studies is the mass yield (grams of material removed per incident ion). We denote this quantity as Y_m, where ϕ is the angle of incidence of the ion beam relative to the surface normal. Sputtering yields have been shown to vary as the cosine of ϕ,[(2)] therefore determinations made at a known angle can be related to the value at normal incidence according to:

$$Y_m^o = Y_m^\phi / \cos(\phi) \qquad (2)$$

In using the QCM for sputtering yield measurements, the time rate of change in mass is determined by measurement of the time rate of change in frequency of oscillation of the QCM bearing the film of interest deposited on the face exposed to ion bombardment. The arrival rate of bombarding ions (R, ions/sec) can be determined from measurement of the target current density (j, A/cm^2) according to:

$$R = \frac{jAN_o}{qF} \qquad (3)$$

where N_o is Avagadro's number, q is the charge on the incident ion and F is the Faraday. Through use of the Sauerbrey equation (eqn (1)) and eqn (3) $Y_m{}^{\phi}$ may be calculated:

$$Y_m{}^{\phi} = \frac{\Delta m}{R\,\Delta t} = \frac{Fq}{jAN_oC}\left(\frac{\Delta f}{\Delta t}\right) \qquad (4)$$

where C is the correction factor of 1.86 determined by Ullevig[1] for the apparatus used here (ϕ = 30.3^o, see experimental).

For the purpose of determining $Y_m{}^{30.3}$ for PPVF, a film of ca 1500 Å thickness was deposited on one face of a QCM crystal. The depposition procedure and apparatus have been described elsewhere.[2] This thickness was estimated by measuring the resonant freuency of the crystal before and after deposition of the PPVF film and calculating the mass of the film according to eqn (1). This mass was converted to a film thickness using an assumed density of 1.5 g/cm^3.) The dependence of $Y_m{}^{30.3}$ as a function of ion dose for this composite is shown in Figure III.5.1 where 1 KeV Ar+ ions were the incident particles.

The behavior of $Y_m{}^{30.3}$ for PPVF is analogous to the behavior for linear organic polymers previously studied by Ullevig,[1] wherein initially $Y_m{}^{30.3}$ rapidly decreases and finally reaches a steady state value ($Y_{m,ss}{}^{30.3}$) of 1.55(±0.38) x 10^{-23} g/ion at 90% confidence. At higher doses (>1 x 10^{17} ions/cm^2) the measured

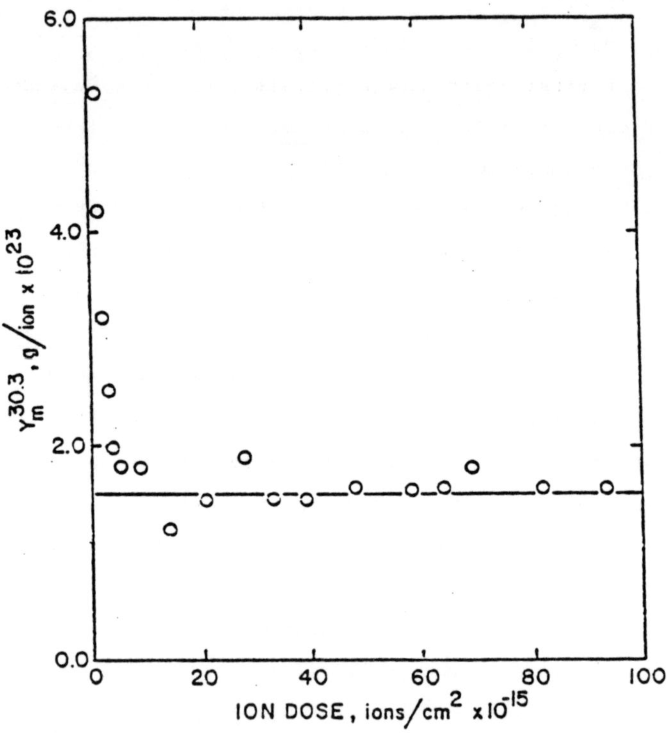

Figure III.5.1: $Y_m^{30.3}$ as a function of ion dose for 1500 Å PPVF film on a Ag Coated QCM crystal. Incident ions were 1 KeV Ar^+. Horizontal line indicates the mean $Y_{m/ss}^{30.3}$.

mass yield increased and Ag species from the underlying electrode dominated the positive ion SIMS spectrum, indicating that the rough interface between PPVF and Ag had been reached at a does of ca 1×10^{17} ions/cm^2.

Expansion of the low dose data (ion dose $<1 \times 10^{15}$ ions/cm^2) with fit to a first order damage process yields the dependence shown in Figure III.5.2. This behavior is to be expected for thin films with high organic content.[1] A steady state sputtering yield is achieved when the rate of damage becomes insignificant relative to the rate of removal (sputtering) of the damaged layer. The slope of this plot(s) is proportional to the damage cross (κ) section for 1 KeV Ar^+ bombardment:

$$\kappa = (2.303)(5\ cm^2/ion)(10^8 Å^2/cm)^2 = 30.9(\pm 4.2) Å^2/ion \qquad (5)$$

at 90% confidence.

It is worthwhile to compare this result to the 1 KeV Ar^+ damage cross sections of linear polymers determined by Ullevig.[1] In that study the sputtering behavior of dip-coated linear polystyrene (PS) and linear polymethylmethacrylate (PMMA) (also carried out using 1 KeV Ar^+) was examined, giving κ values of 15.5 $Å^2$/ion and 15.1 $Å^2$/ion, respectively. The fact that the damage cross sections for these linear polymers are significantly less (a factor of ca 2) than that determined for the highly crosslinked PPVF is significant and merits interpretation.

The most obvious explanation for this difference in damage cross section is based on differences in density which could

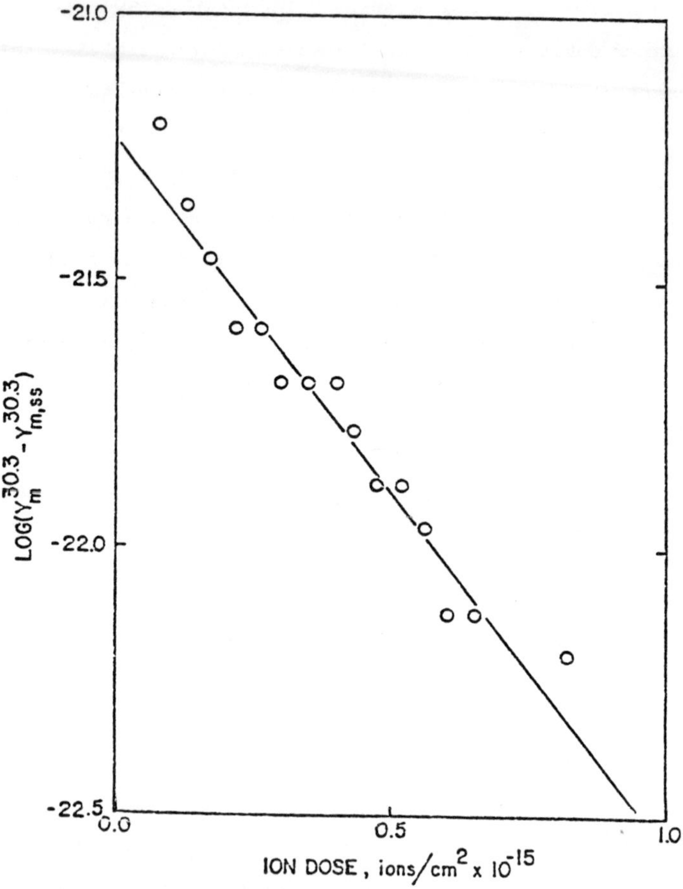

Figure III.5.2.: Log($Y_m^{30.3}$-$Y_{m,ss}^{30.3}$) as a function of ion dose for film of Figure 5-1, showing expansion low dose first order damage process. Solid line is the regression line: slope = $-1.34(\pm 0.17) \times 10^{-15}$ cm^2/ion at 90% confidence, correlation coefficient = 0.9739.

affect the trajectory of the primary ions _within the film_ and the displacements of those atoms undergoing inelastic collisions with the primary ion. The measured damage cross sections are lower for linear PS and PMMA, the densities of which are 1.1 g/cm^3[3] and 1.2 g/cm^2,[4] respectively, while that for PPVF (_vide post_) is _ca_ 1.5 g/cm^3. This implies that the range (average depth of penetration) of the 1 KeV Ar^+ is less in PPVF, and that the tendency for momentum exchange components established normal to the initial primary ion trajectory during the collision cascase[5] are larger in the denser PPVF film.

The Damage Process as Studied by XPS

The XPS survey spectrum of a _ca_ 200 Å thick PPVF film on single crystal Si shown in Figure III.5.3 indicates a relatively high concentration of surface oxygen. This had been observed previously by Dautartas,[6] who also found a strong dependence of both O/C and Fe/C surface composition ratios on the position of a substrate in the plasma reactor. Electrochemical experiments carried out on PPVF/graphite composites indicated the retention of monomer structure in that the cyclic voltammetry indicated waves corresponding to the oxidation and reduction of dicyclopentadienyl iron (II)/(III) film components. Moreover, the charge passed in oxidation or reduction experiments showed that at least 95% of the iron was present in one of these forms. The observed oxygen was

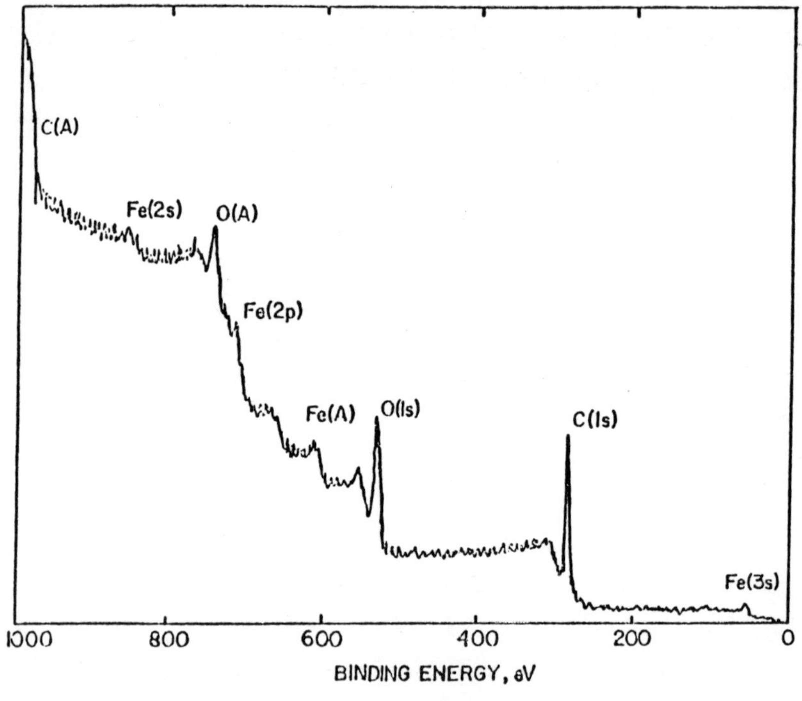

Figure III.5.3.: Low resolution XPS survey scan of ca 200 Å film of PPVF on Si. Pass energy = 200 eV.

therefore concluded to be present either as functional groups on the polymer backbone or as part of a negatively charged counterion present to balance the positive charge of dicyclopentadienyl iron (III) (vide post).

Because of the limited analysis depth of XPS (estimated to be on the order of 50 Å for these samples) data of the type shown in Figure III.5.3 cannot be used to determine whether the PPVF films contan high concentrations of oxygen only at the surface or throughout the bulk of the film. Normally, thin film analysis or depth profiling could answer this question. However, because of the damage incurred by the film during ion bombardment (vide supra) the analytical data obtained with surface sensitive techniques such as the electron spectroscopies (AES or XPS) after ion beam micro sectioning into the bulk must be cautiously interpreted. Such artifacts in the TFA of linear PMMA have been observed by Ullevig.[1] For this material the damage process is found to coincide with preferential sputtering of oxygen during ion beam exposure.

For this reason, it was of interest to use XPS to study the variations in elemental and chemical composition during the early stages of ion beam exposure. The high resolution data shown in Figures III.5.4 through III.5.7 for the $Fe(2p_{3/2})$ region indicate that the ion beam damage process(es) is (are) causing an apparent reduction of Fe (III) to Fe (II). (This reduction process is

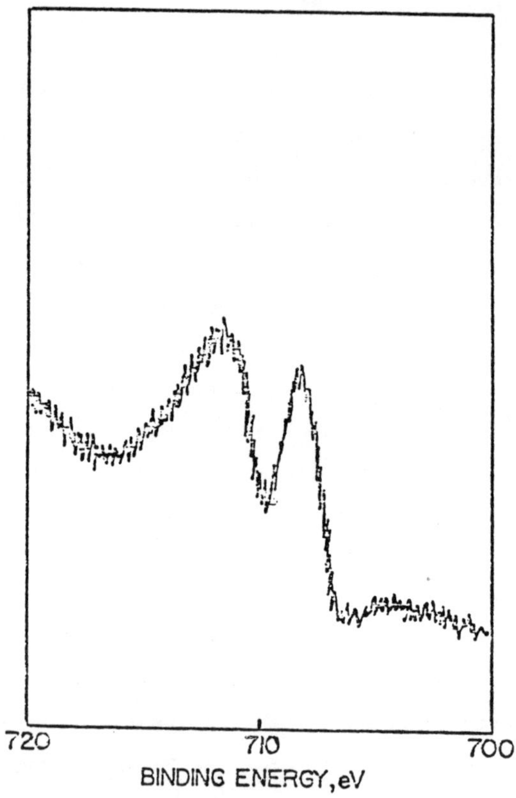

Figure III.5.4: High resolution XPS spectrum of ca 200 Å film of PPVF on Si showing the Fe(2p) region. Pass energy = 25 eV.

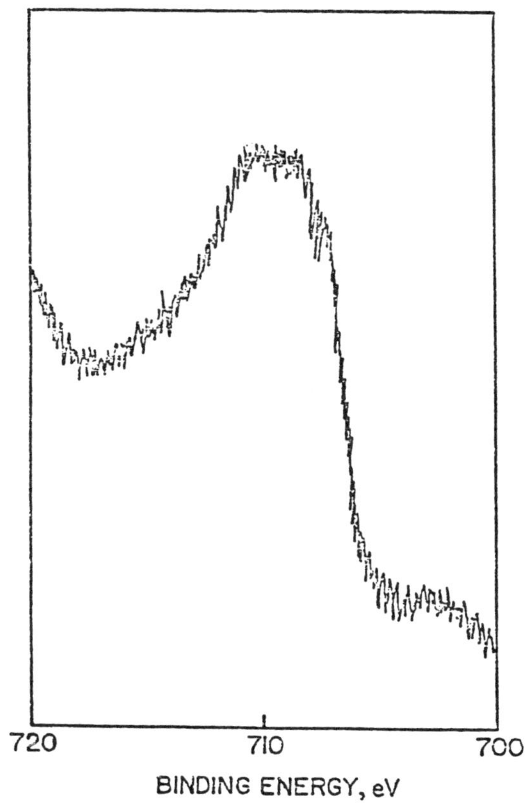

Figure III.5.5: High resolution XPS spectrum of film of Figure 4 after bombardment with a dose of 8.4×10^{14} $ions/cm^2$ (1 KeV Ar^+). Pass energy = 25 eV.

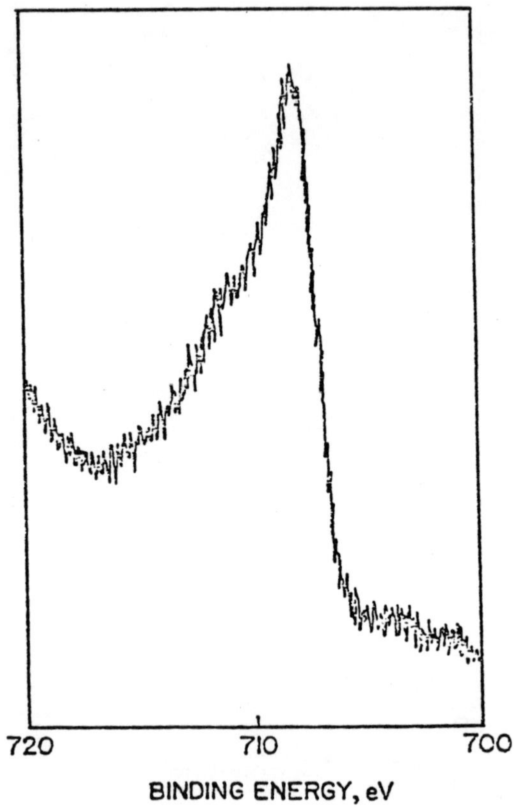

Figure III.5.6: High resolution XPS spectrum of film of Figure 5 after further exposure to 1KeV Ar^+. Total dose = 3.6×10^{15} ions/cm^2. Pass energy = 25 eV.

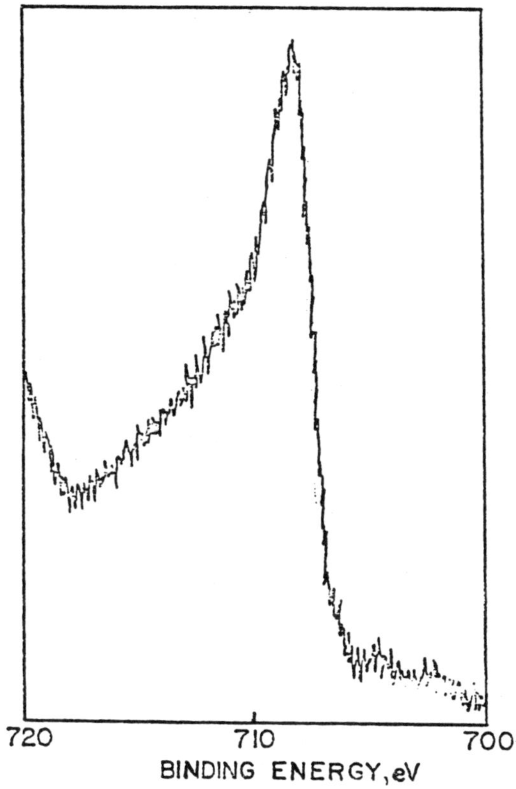

Figure III.5.7: High resolution XPS spectrum of film of Figure 6 after still further exposure to 1 KeV Ar^+. Total dose = 1.2 x 10^{16} ions/cm^2. Pass energy = 25 eV.

reflected in the loss of the Fe (III) peak at 711.6 eV binding energy in favor of the relative increase in Fe (II) peak at 708.3 eV binding energy.) Table III.5.1 shows the variation in Fe/C and O/C elemental ratios (within the analysis volume of XPS) with increased ion dose. The Fe/C ratio is found to increase to a value of 0.080 after an ion dose of 1.2×10^{16} ions/cm^2, which compares favorably with the elemental Fe/C ratio of the vinylferrocene monomer (0.083). Concertedly, the O/C ratio drops by a factor of ca 5.

One is tempted to conclude from these results that the oxygen observed at the surface of the "as deposited" PPVF (see Figure 3) is attributable to the air oxidation of ferrocene ($FeCp_2$) to form ferricenium ($FeCp_2^+$) according to:

$$2\ FeCp_2 + 1/2\ O_2 \longrightarrow 2(FeCp_2^+\quad OH^-) + H_2O \qquad (6)$$

The observation of the Fe (III) peak in the high resolution XPS spectrum prior to ion beam bombardment (Figure 4), and the fact that this species may be electrochemically reduced to Fe (II)‡ supports this hypothesis. Unfortunately, an equally valid conclusion would be that the oxygen is present throughout the film, and is preferentially removed during ion bombardment. Until such time as infrared spectra on the bulk PPVF film become available, the question of the chemical state of the oxygen in the film and its distribution as a function of depth cannot be resolved.

---

‡Preliminary experiments[7] have shown that potentiostating a freshly prepared PPVF/graphite electrode at 0.0 V vs Aq. Ag/AgCl in acetonitrile/0.1 M tetrabutylammonium perchlorate transforms the high resolution XPS $Fe(2p_{3/2})$ spectrum from that shown in Figure 4 to one resembling that shown in Figure III.5.6.

Table III.5.1: Effect of Ion Beam Damage on Composition of Surface Region of PPVF as Determined by XPS

Dose, ions/cm^2	Elemental Composition[a,b]			Elemental Ratios[b]	
	%C	%O	%Fe[c]	Fe/C[c]	O/C
0	77	21	2	0.026	0.273
8.4×10^{14}	83	12	5	0.060	0.145
3.6×10^{15}	87	8	5	0.057	0.092
1.2×10^{16}	88	7	5	0.080	0.057

[a] Excluding H which cannot be detected by XPS

[b] Corrected for elemental sensitivity

[c] Includes all valences of Fe

Comparison of Interface Definition Using AES vs SIMS in the Thin Film Analysis of PPVF on Single Crystal Silicon

Although the depth profiling results discussed above are inconclusive as to the chemical state of oxygen in the bulk vs surface of thse films, thin film analysis does offer a means of determining film thickness. The accuracy of such determinations for opolymer films depends most strongly on two factors. The first is a general consideration (i.e. not restricted to organic or organometallic overlayers) involving selection of a detection technique (SIMS vs AES or XPS) which allows for the most accurate definition of the ion dose required to reach the overlayer/substrate interface. With a knowledge of this value, the sputtering yield (or rate) and primary ion current density the thickness of the film may be calculated. The second factor which must be considered in the determination of the thickness of films which are susceptable to ion beam damage (i.e. organic or organometallic polymers) is the extent to which the initial variation in yield will compromise the accuracy of thickness determinations. In this section the choice of analytical technique will be discussed, and the effects of nonideal sputtering behavior will be deferred to the following section.

In the previous chapter it was conclusively demonstrated that in cases of silane overlayers SIMS is the preferred thin film analytical technique. In general the superiority of SIMS in this regard may be anticipated due to its increased surface sensitivity over the electron spectroscopies. However, this generalization is valid if and only if a suitable species may be identified which can be monitored using SIMS. For purposes of thickness determination the SIMS signal may come either from the film or from the substrate, but preferably from the latter to avoid the effects of knock-on which degrade interface definition.

Several PPVF/Si samples were anlayzed by simultaneous AES/SIMS thin film analysis to ascertain which technique gave the best definition of the PPVF-Si interfact. In Figure III.5.8 the AES/TFA profile of a 680 / thick film of PPVF on Si is shown. The corresponding SIMS profile is given in Figure III.5.9. Clearly, the interface defined in the SIMS profile for $m/e = 28^+$ is much sharper than those in the AES profile.

An interesting aspect of the SIMS profile is that the $m/e = 56^+$ could not be used for thickness determination since no interface is observed. The curve for $m/e = 56^+$ shows that the Fe^+ signal decreases while from AES or XPS measurements the iron content of the film is known to increase during the early stages of sputtering. This is an excellent example of the extremely variable ionization efficiency which has plagued those who have used SIMS to carry out quantitative analyses. Early in the profile where surface oxygen is abundant, iron sputtered from the film is more efficiently ionized (chemical effect). As the oxygen content at the surface region drops, so does the ionization efficiency. At the interface and beyond there are apparently two

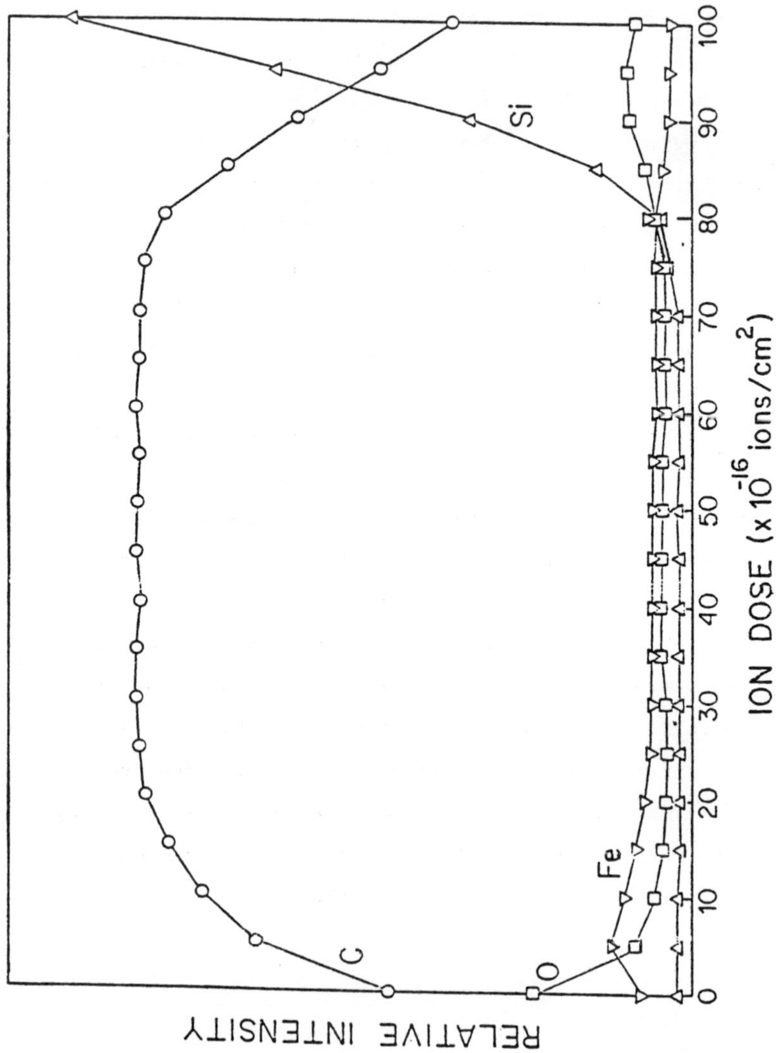

Figure III.5.8: AES thin film analysis of 680 Å thick PPVF layer on Si. Primary ion beam = 1 KeV Ar^+. Primary electron beam energy = 2 KeV.

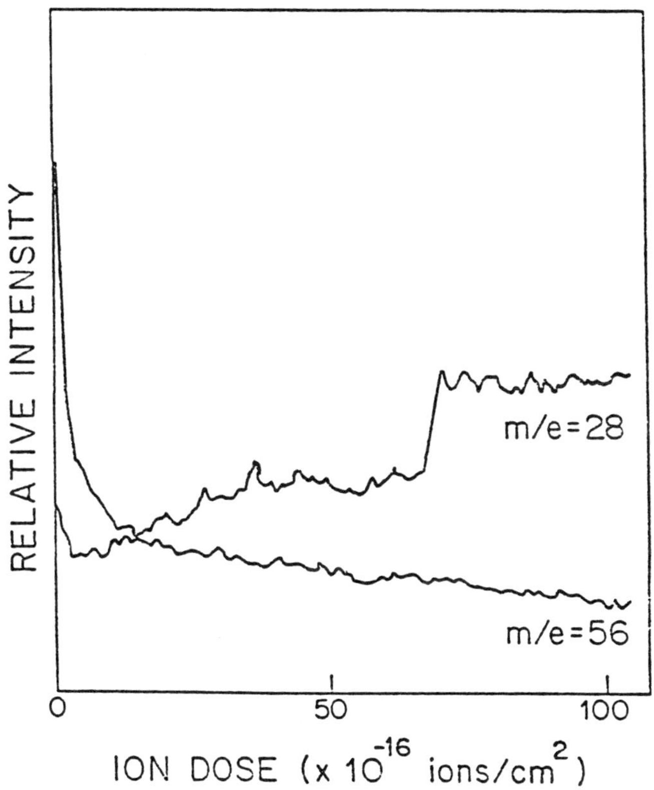

Figure III.5.9: Positive ion SIMS thin film analysis of 680 Å thick PPVF layer on Si. Primary ion beam = 1 KeV Ar^+.

contributions to the m/e = 56^+ signal. The first is from Fe which is knocked into the substrate, and the second arises from the Si_2^+ dimer.

The results shown in Figures III.5.8 and III.5.9 are typical for those samples which were characterized by simultneous AES and SIMS/TFA. For five samples analyzed in this manner (ranging from 680-1150 Å in thickness of the PPVF overlayer) the interface width defined by SIMS was at most half of that defined by AES.[‡] For this reason SIMS was chosen for use in comparison TFA depth determinations to depths determined ellipsometrically.

<u>The Relationship Between Ion Dose Required to Reach the PPVF/Si Interface and Ellipsometrically Determined PPVF Film Thickness</u>

A total of eight samples of PPVF on Si were analyzed first ellipsometrically, and then by SIMS/TFA to determine the reliability of the latter technique in this type of determination. The ellipsometric data given in Table III.5.2 attest not only to the relative uniformity of deposition in terms of film thickness (samples 1 and 2 come from a given synthesis, 3-5 from a second, and 6 and 7 from a third deposition run), but also to the reproducibility of the optical properties from run to run as reflected in the values determined for the index of refraction of the various PPVF films (1.62(±0.08) at 90% confidence).

[‡]Interface width is arbitrarily defined as the ion dose required to cause a change in intensity from 0.7 of that in the overlayer to 0.3 of that in the overlayer. If a substrate signal is monitored then the width is defined as the dose required for a change in intensity from 0.3 of that observed in the bulk of the substrate to 0.7 of that in the bulk.

Table III.5.2: Ellipsometric Data for PPVF Films Deposited on Si Subsrtates

Sample Number	Thickness (Å) Mean	Confidence Limits[ab]	Refractive Index[a] Mean	Confidence Limits[ab]
1	672	662 682	1.646	1.635 1.657
2	676	666 685	1.655	1.645 1.667
3	881	875 887	1.593	1.589 1.598
4	863	857 864	1.600	1.595 1.605
5	858	851 864	1.586	1.581 1.592
6	1265	1262 1267	1.664	1.661 1.667
7	1155	1152 1158	1.660	1.657 1.663
8	2752	3507 3049	1.572	1.474 1.675

[a]No imaginary contribution due to absorption of indicent 632.8 am light

[b]Confidence limits at 95%

Since the ion beam in our apparatus impinges on the sample at a 30.3^{o} angle relative to the surface normal, the depth (d') through which the sputtering of the overlayer must proceed in order to reach the interface is given by:

$$d' = d/\cos(\phi) = d/\cos(30.3^{o}). \qquad (7)$$

A plot of d' values (from correction of the ellipsometric data of

Table III.5.2) vs the ion dose required to reach the PPVF/Si interface is given in Figure III.5.10. The linearity of this relationship is quite good indicating that even in cases where sputtering damage is incurred by the overlayer SIMS/TFA can be reliably used to determine film thickness. Although the regression line shown in Figure 5-12 has a nonzero intercept statistically this line does pass through the origin when a 75% confidence limit in the intercept is considered.

These results appear somewhat contradictory. That is, data of the type shown in Figures III.5.1 and III.5.2 would seem to imply that imprecision is to be expected due to the initially high sputtering yeild especially for the thinner samples. This expectation is valid if one considers the first order damage process for much thinner films than those examined here. Modelling studies carried out by Evans[8] quantitate the anticipated error and for the steady state and initial yield and damage cross section determined for PPVF films the errors listed in Table III.5.3 can be expected for the given film thickness.

Table III.5.3: Errors Predicted for Very Thin PPVF Films Using a First Order Damage Model

Effective Thickness, d', Å	Actual Thickness, d, Å	Dose Required to Reach Interface ions/cm^2	% Error in d'
8	6	6.9×10^{16}	-10%
17	15	1.5×10^{17}	- 5%
44	38	3.8×10^{17}	- 2%
90	77	7.7×10^{17}	- 1%
175	150	1.5×10^{18}	-0.5%

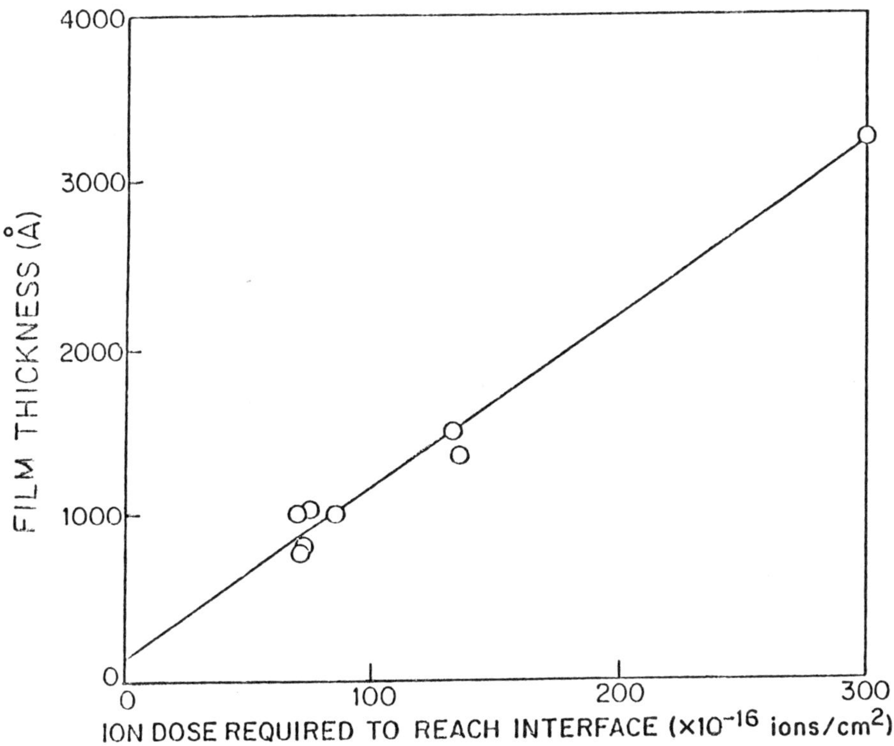

Figure III.5.10: Experimentally observed relationship between effective film thickness (d'), determined ellipsometrically, and ion dose required to reach the PPVF/Si interface by SIMS/TFA. Peak monitored by SIMS was m/e = 28^+. Regression line: slope = 9.95×10^{-24} cm^3/ion;

intercept = 122 Å with standard deviation of 77 Å;

correlation coefficient = 0.9911.

It is evident that systematic errors attributable to the nonideal sputtering behavior of PPVF are insigifniciant (<0.5%) for even the thinnest film studied here (Sample #1, d = 672 Å).

A final film parameter which may be determined from the data obtained here is the density of the PPVF films (δ). δ may be calculated from the slope (s) of Figure 5-12 and $Y_{m,ss}^{30.3}$, according to:

$$\delta = \frac{Y_{m,ss}^{30.3}}{s} \left(10^8 \frac{\text{Å}}{\text{cm}}\right) \qquad (8)$$

Using the statistical uncertainty in $Y_{m,ss}^{30.3}$, one calculates δ to be 1.56(±0.38) g/cm^3 at 90% confidence. By comparison, the density of ferrocene is 1.49 g/cm^2.[9]

The foregoing provides the first evidence that reliable TFA thickness determinations can be achieved for films which sputter nonideally (i.e. show evidence of ion beam damage with concimittant diminution in yield. This study has demonstrated that use of the QCM is definitely required to achieve success in this regard. Moreover, other parameters such as film density are determinate as a natural consequence of carrying out both ellipsometric, TFA and QCM experiments on the same system.

References

1. Ullevig, D. M., Evans, J. F., Anal. Chem., 52, 1467 (1980).
2. Oechsner, H., Appl. Phys., 8, 185 (1975).
3. Bandrup, J., Immergut, E. H., "Polymer Handbook", Interscience; New York, 1966.
4. Miller, M. L., "The Structure of Polymers", Reinhold Publ. Corp.; New York, 1966.
5. Wolf, G. K., "Chemical Effects of Ion Bombardment", Forsh. Chem. Forschung, 85, 1 (1979).
6. Dautartas, M. F., Evans, J. F., "EC Catalysis of Ascorbic Acid Oxidation Using Plasma Polymerized Vinylferrocene Film Electrodes", J. Electroanal. Chem., 109, 301 (1980).
7. Dautartas, M. F., unpublished results.
8. Evans, J. F., unpublished results.
9. Rosenblum, M., "Chemistry of the Iron Group Metallocenes: Ferrocene, Ruthenocene, Osmocene", Interscience; New York, 1965.

III.6. Redox Thermodynamics of PPVF Film Conversion: Comparisons to the Linear Polyvinylferrocene System

In this section we will describe one aspect of the electrochemical characterization of the plasma polymerized vinylferrocene thin films as prepared and analyzed by the procedures described in previous sections of this report. This aspect is the thermodynamic (equilibrium) behavior of the films as reflected in the electrochemical free energy changes undergone by the films as they are potentiostatically converted from one redox form to another. In the first part of this section the focus is on PPVF films. For reasons which will become apparent, we felt that it was critical to also examine (on a comparative basis) the redox thermodynamics of linear polyvinylferrocene (LPVF) films as prepared by spin coating. As such the structural dissimilarities of the polymer systems are readily inferred from their vastly different thermodynamic behavior. The latter part of this section is devoted to this comparison.

Redox Thermodynamics of PPVF Film Conversion

In the foregoing sections it was established that highly crosslinked PPVF films could be reproducibly synthesized with film properties (especially film density) independent of deposition time. This allowed for systematic evaluation of the thermodynamics of PPVF films of different thickness in various solvent-supporting electrolytes.

It was noted earlier that the $E^{o'}_{app}$ for PPVF films, made in the prototype reactor, were invariant when measured in 0.1 M TBAP/CH_3CN and 0.1 M NaCl/H_2O (pH 2.5 glycine buffer). In contrast, the $E^{o'}$ for solubilized ferrocenes was shifted by 200 mV in these respective solvent - supporting electrolytes. It was stated that the invariance in the $E^{o'}_{app}$ for the PPVF films might be a consequence of ferrocene centers in the film experiencing the same environment, the polymer film itself, in both solvents.

The dependence (or lack thereof) of the $E^{o'}_{app}$ for PPVF films on electrolyte media was further evaluated by expanding the number of solvent - supporting electrolytes (SSE) examined. The results of this investigation are given in Table III.6.1. Hydroxymethylferrocene (HMF) was used for comparison of PPVF results to solubilized ferrocenes because of its greater solubility in water than ferrocene or vinylferrocene.

It is readily apparent that PPVF $E^{o'}_{app}$ of PPVF appears to be much less dependent on the SSE than the $E^{o'}_{app}$ of HMF. Because of differences in liquid junction potentials between different SSE systems, it is not known if the variation in the PPVF $E^{o'}$ is simply a manifestation of liquid junction potential variation. Of course, this variation would also occur to the same extent in the case of measured $E^{o'}$s for HMF.

Table III.6.1: Apparent $E^{o'}$ for PPVF in various electrolytes.

SSE	$E^{o'}_{app}$,[a] V vs. Ag/AgCl	
	PPVF[b]	HMF[c]
0.1 M TBAP/CH_3CN	0.395	0.390
0.1 M TEApTs/CH_3CN	0.404	0.372
0.1 M $TBAPF_6$/CH_3CN	0.392	0.369
0.1 M TBAP/CH_2Cl_2	0.430	0.467[d]
0.1 M $LiClO_4$/H_2O	0.304	0.192
0.1 M KCl/H_2O	0.421	0.240
0.1 M KNO_3/H_2O	0.380	0.240
0.1 M KF/H_2O	0.429	0.245

a) Measured by slow scan cyclic voltammetry, $\nu = 10\ mV\text{-}s^{-1}$, $E^{o'}_{app} = \frac{E_{p_a} + E_{p_c}}{2}$

b) All PPVF films between 750 - 800 Å

c) ΔE_p in all cases was between 56 and 60 mV at $\nu = 10\ mV\text{-}s^{-1}$ [HMF] = ca 0.2 mM

d) $\Delta E_p = 77$ mV at 20 $mV\text{-}s^{-1}$

That the PPVF film is poorly solvated is a fair assumption, considering the concentration of ferrocene in the film (4.38 M). For comparison, the concentration of ferrocene in crystaline ferrocene is 8.05 M and the concentration of ferrocene in linear PVF is 5.89 M. Since the PPVF film is highly crosslinked and rather dense, having most likely a small free volume in the film, in the absence of PPVF film swelling it would be difficult to accommodate much solvent in the PPVF film, and thus solvation effects on the free energy of redox should be far less important than that for solubilized ferrocenes in solution. If it is assumed that the $E^{o'}$ of PPVF is unaffected by the SSE, due to poor solvation, and furthermore assumed that the PPVF film examined in KCl/H_2O is least affected by liquid junction potential between the SSE and the reference electrode (Ag/AgCl 1.0 M KCl), then the difference between PPVF $E^{o'}$ in other SSE is a measurement of the liquid junction potential.

We emphasize that the above discussion is speculative at this point. However, we note the possible invariance of PPVF $E^{o'}$ with SSE because of rather significant ramifications of this invariance. PPVF film electrodes (and possibly other electroactive immobilized polymers) might be used as universal reference electrodes, free of junction potential since they might be used in the same SSE as the redox couple of interest in solution. The PPVF reference electrode (or other polymer reference electrode) would most likely require separation from the test solution (e.g. via a frit) to prevent redox between the PPVF reference electrode and the redox couple in solution. The PPVF reference electrode might then be used to examine solvent effects on the redox couple in solution, free of junction potential error.

Pierce and Bard[1] have demonstrated that PVF adsorbed on Pt can be used as relatively stable reference electrodes in acetonitrile. However, in other solvents the PVF/Pt ppotential drifted which was attributed to dissolution of the polymer. They suggested that higher molecular weight and less soluble polymers would extend the range of solvents in which such reference electrodes might be used. PPVF is such a polymer. Clearly, further investigation is warranted in the application of electroactive polymer modified electrodes as potential universal reference electrodes.

In the electrochemical investigations of PPVF electrodes made in the prototype reactor, it was noted that the cyclic voltammetric waves had FWHM > 90 mV, and this was attributed to repulsive interactions between electroactive ferrocene centers in the PPVF films. Larger FWHM in the cyclic voltammograms of PPVF in water were found than in the case of PPVF in acetonitrile. This suggested that interactions between electroactive sites were greater in water, but the results were inconclusive due to the slower kinetics of PPVF redox in water, which might have distorted the voltammetric curves. Because of these possible kinetic artifacts, the potential - concentration relationships of PPVF films were reevaluated (with films made in the new reactor) using the double potential step coulometric method (described in Chapters 5 and 6) in order to ensure the measurements were made at equilibrium. Charge ($Q_a(E)$) due to conversion of ferrocene to ferricenium was measured at 50 mV intervals in a 0.9 V potential window centered at the $E^{o'}_{app}$ of PPVF in the SSE of interest.

Figure III.6.1 shows the charge-potential relationship ($Q_a(E)$ vs. E) for a 749 Å PPVF film on glassy carbon in 0.1 M TBAP/CH_3CN. The surface coverage (Γ_T) for this electrode (measured in 0.1 M TBAP/CH_3CN) was 3.28 x 10^{-8} moles-cm^{-2} (Γ_T = nFAQ_T, Q_T = $Q_a(E)$ at E = 0.9 V vs. Ag/AgCl), whereby the total concentration of electroactive ferrocene sites in the polymer film is 438 M. Note the S shaped curve characteristic of those for potentiometric titrations.

Figure III.6.2 shows the plot of log (ferricenium to ferrocene) as a function of potential (determined via E vs. log($Q_a(E)/Q_T-Q_a(E)$)). The plot is linear between 0.15 volts and 0.45 volts, in which the ferricenium/ferrocene concentration ratio changes by ca 3 orders of magnitude. The slope of this linear region was 126 mV with an intercept of 0.421 V. However, the plot deviates from a straight line at potentials anodic of 0.45 volts. There are several factors that might contribute to this deviation from linearity.

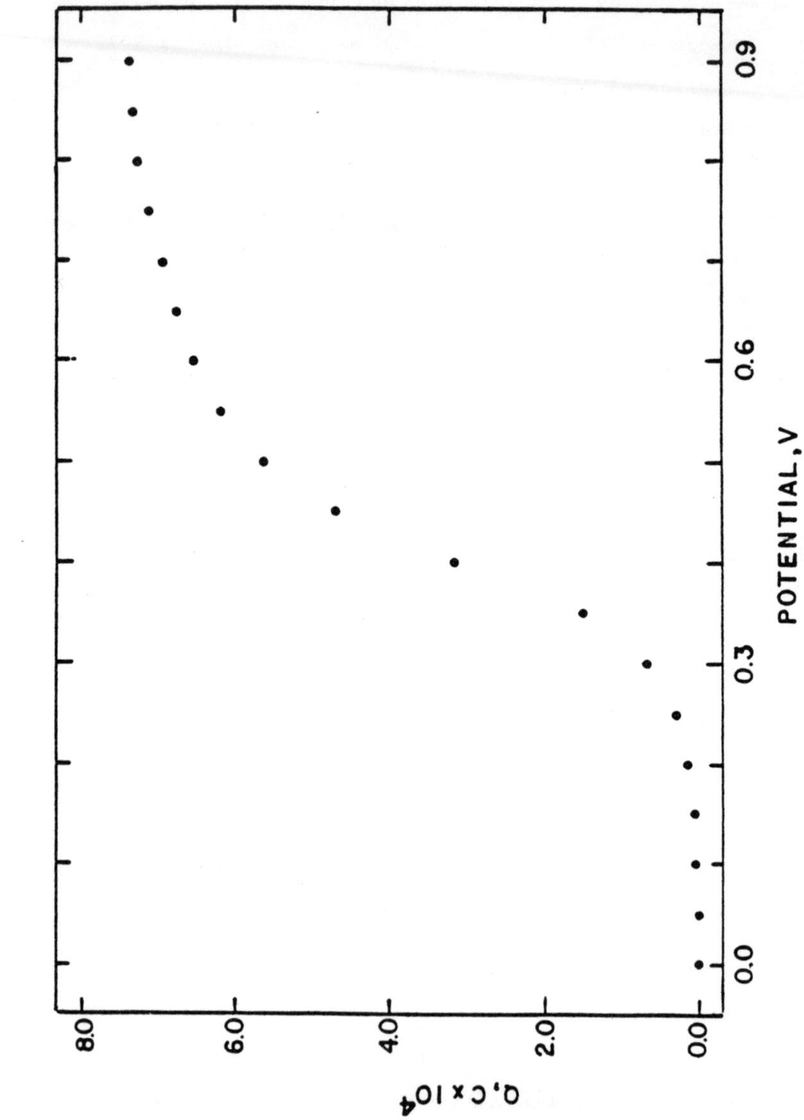

Figure III.6.1: $Q_a(E)$ as a function of E for a 750 Å PPVF film on glassy carbon in 0.1 M TBAP/CH_3CN.

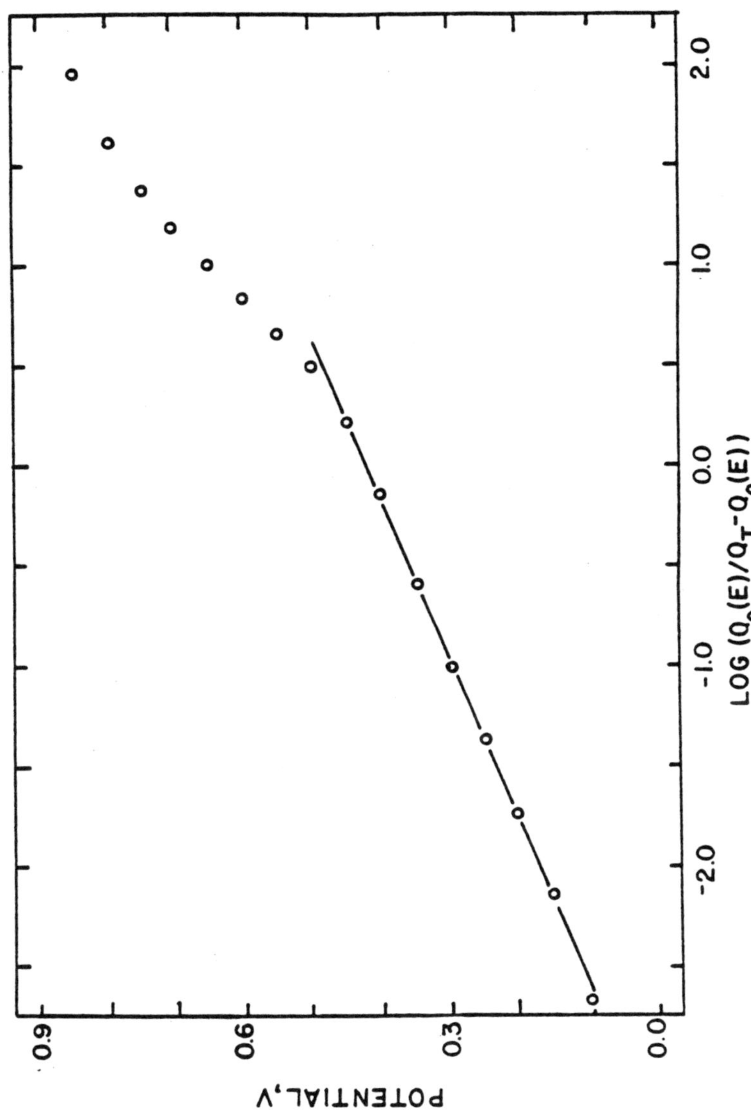

Figure III.6.2: E vs. log $\frac{Q_a(E)}{Q_T-Q_a(E)}$ for 750 Å PPVF film on glassy carbon in 0.1 M TBAP/CH_3CN.

Although the PPVF film was quite stable in 0.1 M TBAP/CH_3CN, some decomposition occured. This decomposition was most pronounced at potentials anodic of the $E^{o'}$ of PPVF. For example, in duplicate potential steps from 0.0 to 0.9 volts, the charge accumulated on the second potential step ws typically 0.25% lower than that of the first step. This decomposition would cause the charge measured for subsequent potential steps cathodic of 0.9 V to be underestimated and give rise to at least part of the deviation from linearity at potentials anodic of 0.45 volts. However, this would only occur for experiments wherre the charge as a function of potential was measured with the initial step from 0.0 V to 0.9 V to 0.0 V, and subsequent steps were from 0.0 V to 0.85 V to 0.0 V, 0.0 V to 0.80 V to 0.0 V, etc. By reversing the order of the potential steps (e.g. 0.0 V to 0.1 V to 0.0 V, 0.0 V to 0.15 V to 0.0 V, etc.) the effect of decomposition on the charge - potential plots should be minimized. This was evaluated for experiments conducted in both directions, and no appreciable difference was found in the calculated slope and intercepts, and in the deviation from linearity at potentials anodic of 0.45 V. Furthermore, the difference between the charge measured at 0.9 V and 0.85 V was -2.5%, 10 times greater than the typical amount of charge lost as a consequence of decomposition in replicate potential steps from (0 to 0.9 V). Hence, the contribution to the nonlinearity as a consequence of decomposition was judged to be a minor contribution.

Another factor that may contribute to the deviation from linearity might be a consequence of not accounting for charge due to capacitance. The effect of capacitance on coulometric errors was discussed in section III.1 and was shown to be relatively minor (<1.5%) for surface coverages greater than 5 x 10^{-9} moles-cm^{-2}, assuming the electroactive modifier had a Nernstian response (60 mV per decade change in O/R ratio at 25°C) and that the electroactive polymer does not contribute to the capacitance. The PPVF films do not have a simple Nernstian response as indicated by the 126 mV per decade change in concentration ratio, and hence a wider potential window was required for the charge potential - experiment (900 mV). Using a 900 mV window and the same assumptions given in section III.1 for estimating surface coverage determination (Q_{dl} = 20μf), the surface coverage of a film whose actual coverage was 5 x 10^{-9} moles-cm^{-2} would be overestimated by 5%. This is an appreciable error and could give rise to some of the deviation from linearity. However, since the capacitance of the film covered electrode was not known, no attempts were made to correct for this possible source of error.

The existence of a higher potential couple, observed in the cyclic voltammetry of PPVF film electrodes may also give rise to the deviation from linearity at potentials anodic of 0.45 volts.

It was estimated this high potential couple comprises _ca_ 10% of the total concentration of electroactive ferrocene sites in the PPVF films. The effect of a higher potential couple on potential-charge relationships was evaluated by calculating the expected relationship for the following assumptions:

1) $E^{o'}$ of the low potential couple = 0.40 V;

2) $E^{o'}$ of the higher potential couple = 0.60 V;

3) the two couples have a log concentration response _vs._ E which is 120 mV/decade.

The calculated charge - potential relationships for 0, 10, and 20% of the total redox sites comprised of the higher potential couple are shown in Figure III.6.3, and the log of the concentration ratio _vs._ E is shown in Figure III.6.4. Qualitatively, Figure III.6.3 resembles Figure III.6.1. Comparing Figure III.6.4 to Figure III.6.2, the deviation from linearity occurs in the same region for the model as seen for the PPVF film. However, the correspondence is not exact, most probably due to the decomposition mentioned previously and also due to the fact that at potentials anodic of 0.65 volts, small errors in measured $Q_a(E)$ produce large deviations in the log concentration _vs._ E plot. This is readily seen by examining Figure III.6.3 in that the difference in $Q_a(E)$ for potentials anodic of 0.65 volts are small with respect to Q_T.

In general, only points between 0.15 and 0.45 volts fell on a straight line in log concentration ratio _vs._ E plots for PPVF films. What is of importance here then is the effect of the higher potential couple on the calculated slope and intercept using only the points between 0.15 and 0.45 volts. These results (using the above model) are given in Table III.6.2.

Table III.6.2 Effect of a higher potential couple on potential log-concentration relationships.

% higher potential couple	slope (mV)	Intercept (V vs. Ag/AgCl)
0	120	0.40
10	124	0.412
20	130	0.425

It is noted that the intercept calculated is more affected by the presence of the higher potential couple than is the slope. Based on the above mode, and comparison to the results for the 749 Å PPVF film (Figure III.6.2), the fractional amount of the film comprised of the higher potential is estimated to be 13%. However, this assumes the higher potential couple $E^{o'}$ is 0.6 V. In principle, if this $E^{o'}$ was known, the curves could be fit to calculate the relative concentration of both couples. Since the anodic wave of this couple was not observed, an accurate $E^{o'}$ for this couple could not be assigned. Hence, the slopes and intercepts are calculated from the log concentration ratio - potential plots using the points from 0.15 to 0.45 V (or the equivalent window relative to $E^{o'}_{app}$) without correcting for the

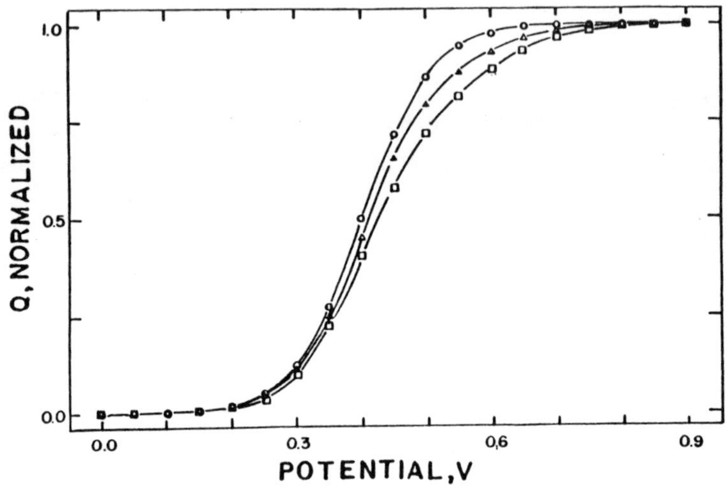

Figure III.6.3: $Q_a(E)$ vs. E for an electroactive polymer containing two electroactive couples with different $E^{o\prime}$'s.

○ 100% of electroactive sites with $E^{o\prime} = 0.40$ V

△ 90% of electroactive sites with $E^{o\prime} = 0.40$ V, 10% with $E^{o\prime} = 0.60$ V.

□ 80% of electroactive sites with $E^{o\prime} = 0.40$ V, 20% with $E^{o\prime} = 0.60$ V.

Both electroactive couples have a 120 mV response per decade change in concentration ratio of oxidized to reduced form of the couple.

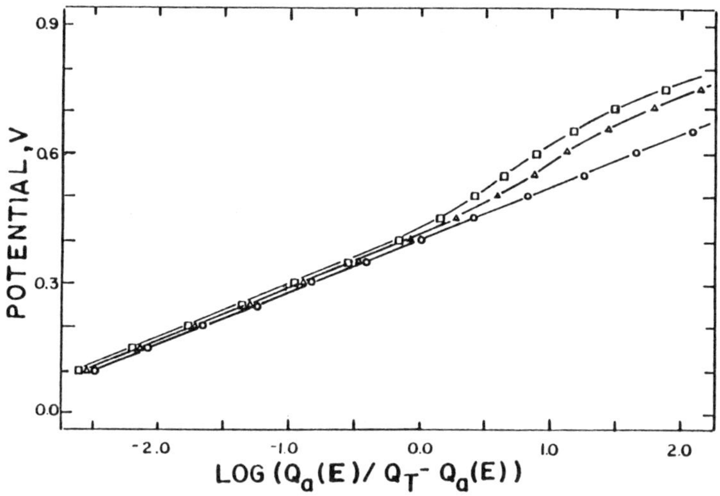

Figure III.6.4: E vs. $\frac{Q_a(E)}{Q_T - Q_a(E)}$ for model polymer of Figure 3.

○ 100% of electroactive sites with $E^{o\prime}$ = 0.40 V.

△ 90% of electroactive sites with $E^{o\prime}$ = 0.40 V, 10% with $E^{o\prime}$ = 0.60 V.

□ 80% of electroactive sites with $E^{o\prime}$ = 0.40 V, 20% with $E^{o\prime}$ = 0.60 V.

error in the slope and intercept caused by the presence of the higher potential couple.

Irrespective of these errors, the slope calculated from the 752 Å PPVF films in 0.1 M TBAP/CH_3CN is very different from that predicted from the ideal Nernst equation. The above slope was found to be relatively invariant for PPVF films of various thickness, in various solvent - supporting electrolytes, for PPVF films on different substrates, and films of different density. These results are summarized in Table III.6.3. The average slope from all the experiments is 129 ± 9 mV, demonstrating the invariance to the parameters studied here. Clearly, in all cases, the slope is not that expected for ideal Nernstian behavior.

Several models were examined as to what process or processes can account for the slope of the PPVF charge - potential relationship. These included dimerization of $FeCp_2^+$ and partitioning of the counterion in the PPVF film. Both of these, however, do not give rise to a linear relationship between the log concentration ratio to the potential. If the number of electrons transferred was equal to 1/2, this would give a close to Nernstian slope, but this is totally unreasonable. If the Nernst equation is written as below:

$$E = E^{o'} + p\,\frac{0.059}{n}\log\frac{[FeCp_2^{+}]}{[FeCp_2]} \qquad (1)$$

Table III.6.3: Charge - Potential Relationship for PPVF.

Substrate	SSE	d, Å	S, mV/decade	int, V	$[FeCp_2]$, M	Γ moles-cm^{-2}
GC	0.1 M TBAP/CH_3CN	195	149	0.428	4.45	0.863×10^{-8}
GC	0.1 M TBAP/CH_3CN	455	126	0.424	4.27	2.02×10^{-8}
GC	0.1 M TBAP/CH_3CN	750	126	0.421	4.38	3.28×10^{-8}
GC	0.1 M TBAP/CH_3CN	1260	120	0.431	4.52	5.7×10^{-8}
PG	0.1 M TBAP/CH_3CN	--	126	0.426	--	1.32×10^{-8}
PG	0.1 M TBAP/CH_3CN	--	133	0.430	--	2.1×10^{-8}
Pt	0.1 M TBAP/CH_3CN	--	129	0.429	--	1.73×10^{-8}
GC	0.1 M TBAP/CH_3CN	777	143	0.432	2.59[a]	2.0×10^{-8}
GC	0.1 M TEApTs/CH_3CN	761	123	0.404	3.82[b]	2.91×10^{-8}
GC	0.1 M TBAP/CH_2Cl_2	765	121	0.461	3.65[b]	2.79×10^{-8}
GC	0.1 M $LiClO_4/H_2O$	776	125	0.306	3.98[b]	3.09×10^{-8}

(a) nf = 1.52; made in 15.0 mtorr steady state pressure discharge (30 minutes).

(b) These concentrations were calculated from the total surface coverage measured in this SSE.

then, since n = 1, this implies that $p \cong 2$. p is defined here as a nonideality parameter. If this is expressed in terms of activities, where

$$a_{FeCp_2^+} = [FeCp_2^+]^p \qquad (2)$$

$$a_{FeCp_2} = [FeCp_2]^p \qquad (3)$$

the implication is that the activity of ferrocene and ferricenium in the PPVF films is proportional to the square of their respective concentrations. Although the above equations fit the charge - potential response for PPVF films, the explanations for this fit are not obvious.

Solubilized ferrocenes are known to give one-electron transfer Nernstian responses. Furthermore, linear ferrocene polymers in solution also behave according to the Nernst equation (slope = 63 mV/decade)[2]. Bard has examined PVF adsorbed on platinum, and some evidence is given for interaction between electroactive sites[3]. However, the interaction is an attractive one, causing the cyclic-voltammetric peak width to be less than theoretical (40 mV _vs._ 90 mV at half height). If this system was truly at equilibrium, this would indicate a log concentration ratio _vs._ E slope of _less_ than 59 mV/decade. However, no proof of the attainment of equilibrium on the cyclic voltammetric time scale was given.

Murray has examined activity relationship of PPVF film electrodes[4]. He finds the activity relationships depend on solvent and surface coverage of PPVF. Attractive interactions are found in water, repulsive interactions are found in acetonitrile. There results seem to contradict the results given here, since virtually no dependence on solvent was found. Two factors may account for this discrepancy: (1) Murray used iinetic methods (e.g. cyclic voltammetry) to evaluate thermodynamic properties. That the experiment was not conducted at equilibrium is evidenced by the peak separation in the cyclic voltammetric waves. (2) Murray has reported that PPVF films made in his lab are soluble and in many respects behave like linear PVF (this was established by comparison of cyclic voltammetric responses[5].

If the latter is the source of discrepancy between the observed behavior of PPVF studied in this dissertation and PPVF studied by Murray, the implication is that the thermodynamics of polymer film redox depends on polymer film properties (e.g., crosslinking and solubility). Preliminary results on the equilibrium - charge potential relationship for PVF adsorbed on glassy carbon support this hypotehsis, since a slope of 76 mV/decade was found[6]. Further speculation cannot be made until comparisons between PPVF and other electroactive polymers as studied by equilibrium charge - potential relationships are made.

Redox Thermodynamics of LPVF Film Conversion

A cyclic voltammogram (1 mV/s) of a thin LPVF film in contact with 0.10 M TBAP/acetonitrile is shown in Figure III.6.5. The morphology of this curve is very similar to previously reported results[7,8]. An interesting aspect of these voltammograms is the persistence of a separatism between the anodic and cathodic peak potential, ΔEp been at slow scan rates[8]. At $\gamma = 1$ mV/s we find Δ Ep = 15 mV, increasing as a strong function of scan rate to 135 mV at $\gamma = 100$ mV/s for the system of Figure III.6.5. Clearly, γ must be much less than 1 mV/s to anticipate that equilibrium between the LPVF film and the glassy carbon supporting electrode could be established on the time scale of the cyclic voltammetric experiment.

Furthermore, some measure of hysterisis is clearly evident in the data of Figure III.6.5. This dictates the necessity of using coulometric titration procedures to evaluate what the equilibrium thermodynamic behavior of the polymer film is. The coulometric procedure called for incrementing the potential from 0.0 V at which the film is fully reduced to more positive values. If, however, the initial ppotential is chosen to be +0.9 V with successive decrements toward 0.0 V, only the coulometric results obtained at potentials $E>E^{o'}$ are the same as for the incremental procedure. For potentials $E<E^{o'}$ the results do not coincide, indicating that even with the coulometric approach the problem

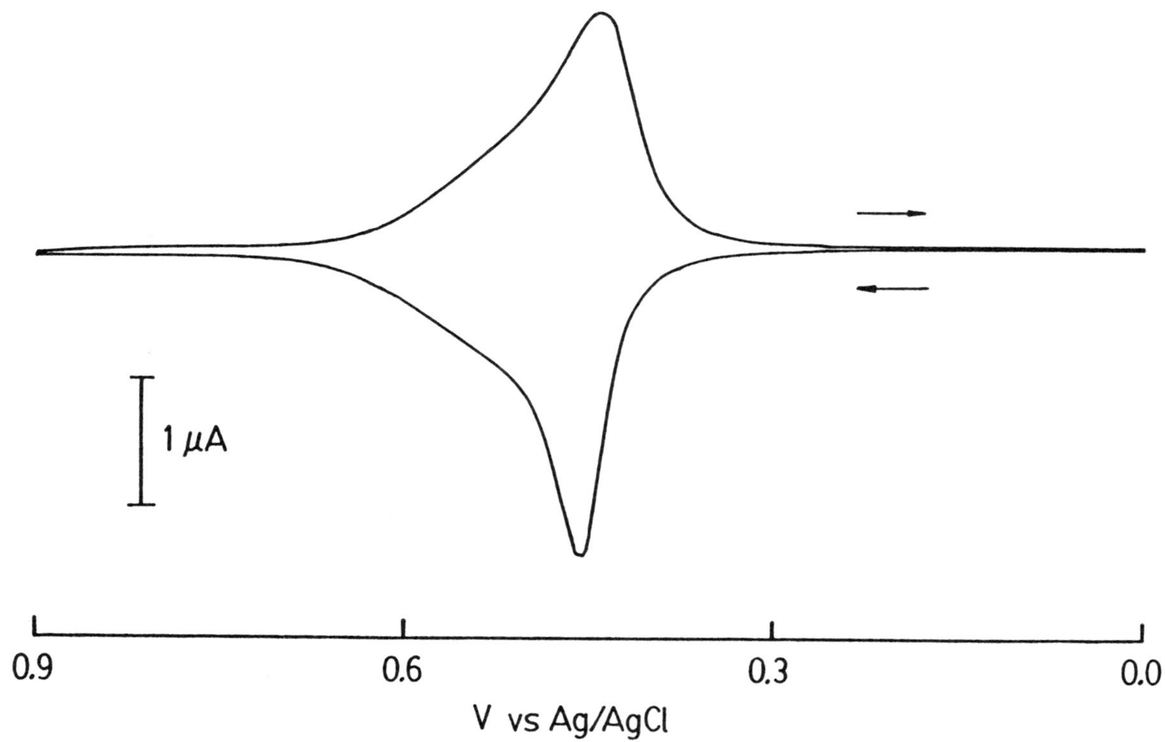

Figure III.6.5: Cyclic Voltammetry of LPVF supported on glassy carbon. Film thickness: 190 Å, SSE: 0.10 M TBAP/acetonitrile, γ = 1.0 mV/s.

slow conversion kinetics cannot be completely overcome. This type of behavior and our interpretation of its causes are discussed below in a later subsection of the results have been presented.

Figure III.6.6 shows a typical titration curve for an LPVF film on glassy carbon. Although this data was taken on a film of 854 Å thickness, it was found that the shape of the titration curves were thickness independent over the range 200-1200 Å. Numerical evaluation of the derivative, df/dE, of the titration curve gives the result plotted in Figure III.6.7. This curve can be thought of as the limit of the cyclic voltammetric response as $\gamma \rightarrow 0$. Because the sequence and time scale of the measurements resulted in practical equilibrium within the film, Figure III.6.7 is representative of the reversible behavior of this system. The shape of this curve is essentially identical to the anodic peak of Figure III.6.5, but is obviously narrower than the cathodic wave. This difference is related to the hysteresis discussed above. This phenomenon will be discussed in detail below.

A typical Nernst plot for LPVF is given in Figure III.6.8. In contrast to a simple, single electron transfer process (linear plot with 59 mV/decade slope at 25°C) the thermodynamics as reflected in this data are quite complex. As a fully reduced LPVF

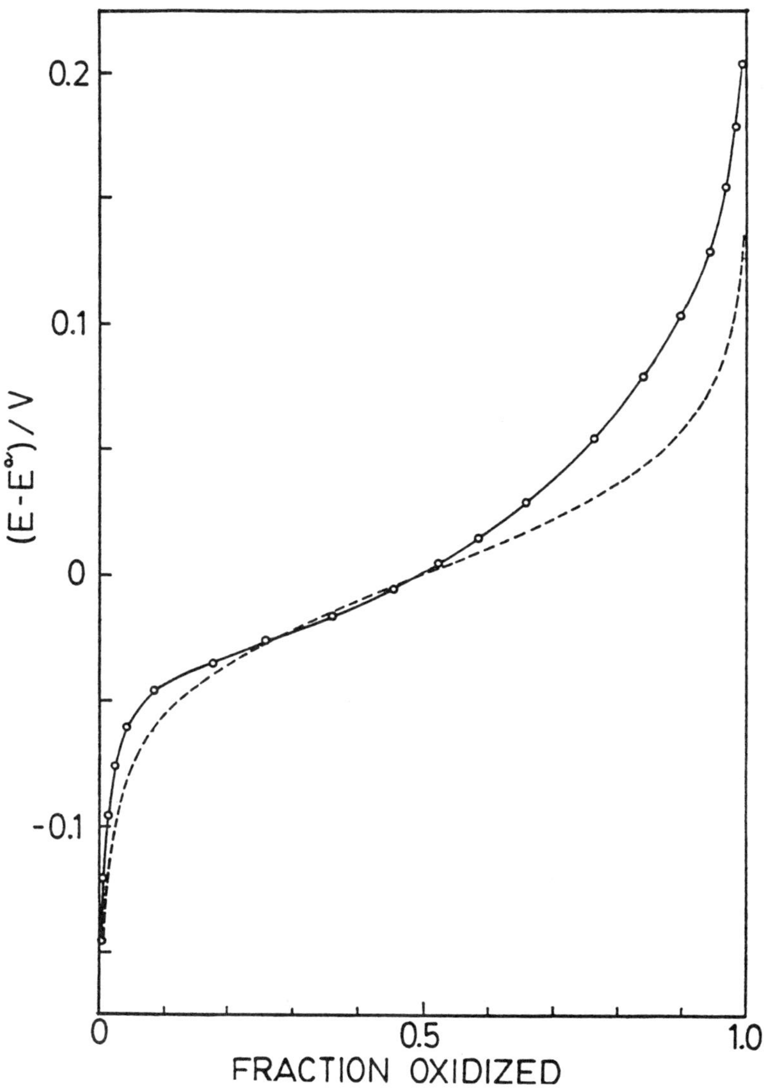

Figure III.6.6: Coulometric titration curve obtained under controlled potential conditions for an 854 Å film of LPVF supported on glassy carbon. Determination carried out in 0.1 M TBAP/acetonitrile. Dotted line is the theoretical curve for a simple one electron couple of the same $E^{o'}$ as the LPVF film. f_{ox} is the fraction of ferrocene sites in the oxidized form.

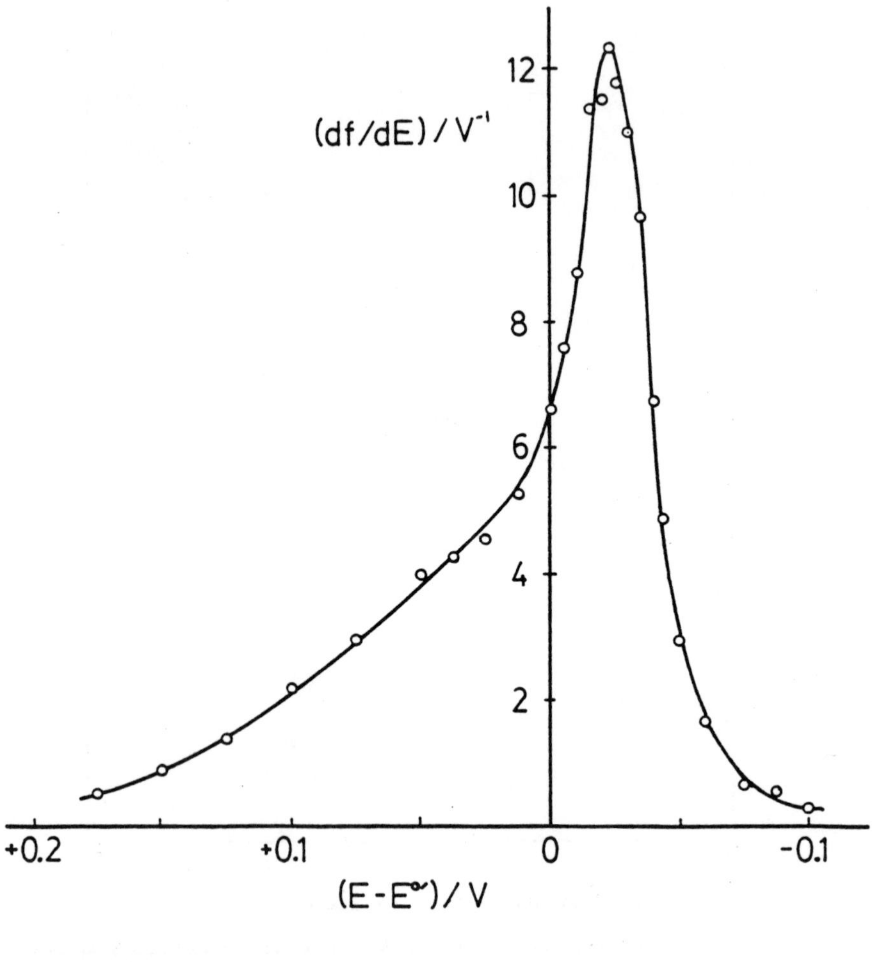

Figure III.6.7: Derivative of the titration curve shown in Figure 6.

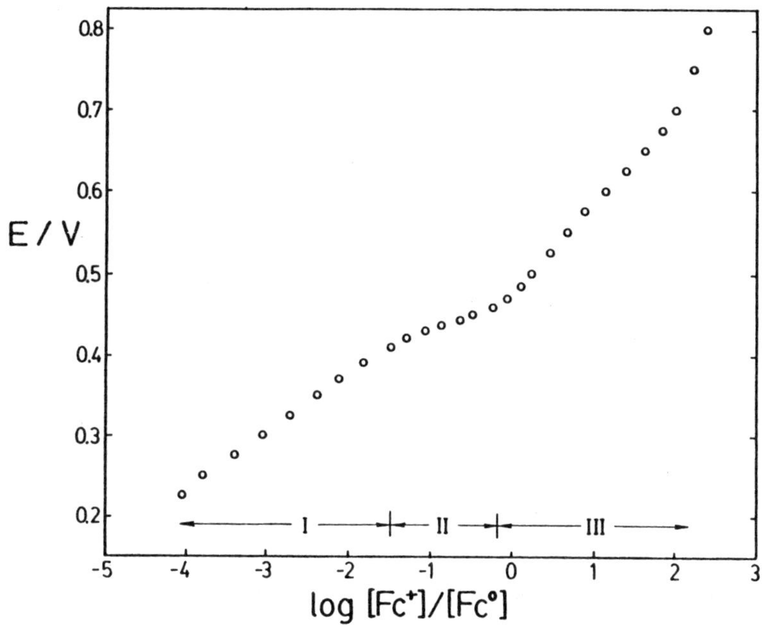

Figure III.6.8: Nernst plot for coulometric titration of LPVF supported on glassy carbon. Data obtained in 0.10 M TBAP/acetonitrile for a 756 Å thick film. See text for a discussion of regions I, II and III.

film begins to undergo oxidation, the curve is approximately linear with a slope of ca 80-100 mV/decade. As the apparent $E^{o'}$ (f=0.5) is approached, the slope decreases, smoothly undergoing a transition to a value of ca 20-25 mV/decade at potentials just cathodic of the $E^{o'}$. At fractional oxidation in excess of 0.5 ($E \geqslant E^{o'}$) the slope rapidly increases to a value of ca 110-120 mV/decade, which is more or less maintained until reachimg the fully oxidized state.

There appear to be three distinct regions of differeing overall behavior. These are labelled as regions I, II and III in Figure 8 where, in the simplest limit, one could assume that the differences in the three regions are attributable to three different primary phenomena which complicate the thermodynamics beyond the expected behavior of the simple case. It should be noted, however, that the potential for the proposed demoscation, are somewhat arbitrary, and clearly these are transition regions wherein there are a multiplicity of complicating phenomena which are simultaneously operative.

The following subsections deal in turn with interpretation of these results on a region by region basis under the assumption that one, primary phenomenon is responsible for the nonideal behavior observed. The first subsection addresses the behavior in region II in the context of a model based on aggregation of ionic redox sites. Region III is then treated in which it is proposed that a mechanical/electrochemical phenomenon may account for the super Nernstian slope. Finally, the response in Region I is discussed in terms of a fixed site ion exchanger in which the potential of the phase containing these sites (ferricinium pendant groups) is governed not only by the ferrocene/ferricinium redox couple, but also by membrane potential contribution from supporting electrolyte ions whose activities and mobilities are different in the polymer and solution phases.

Region II: Redox Site Aggregation Model

The sub-Nernstian slopes observed for Region II of the LPVF Nernst plots are interpreted to indicate some form of site aggregation within the films. Consideration of the literature on ionic polymers [9-11] leads us to the conclusion that site aggregation should be expected to occur in LPVF polymers as they become oxidized. Dry, reduced LPVF is an amorphous, non-ionic, hydrophobic material. Its dielectric constant is most likely in the range of 2-4 by analogy to other hydrophobic polymers[12].

Upon partial oxidation, ferricinium cations and an equal number of counter-ions will reside within the film. Because of the low dielectric constant these ions are not expected to exist as separate ions but rather as ion pairs. This general situation provided the starting point for Eisenberg's theory[13] of ion clustering in organic polymers. This theory predicts that ions will associate to form ion/counterion pairs, that ion pairs can associate into higher order multiplets, and that multiplets can further associate into larger clusters of sizeable dimensions, i.e., 50-100 Å. This last phenomenon is referred to as microphse separation, since the polymer is pictured as being comprised of ionic domains separated by essentially nonionic regions. The nature of association depends on the ion concentration (low values favor lower order multiplets, higher values favor clustering) as well as certain physical properties of the polymer. Results of investigations addressing the effects of ion incorporation on polymer structure are generally supportive of Eisenberg's theory[9-11].

Quantitative application of the theory to LPVF films operating in an electrochemical environment does not seem feasible at present. An estimate of cluster decomposition temperature is needed to perform the calculations, and there is no experimental evidence available on which to base such a value for LPVF. Furthermore, the situation inside an LPVF film exposed to an electrolyte environment is considerably more complex than that for the interior of a dry film. Solvent and electrolyte undoubtedly partition to a certain extent into the film, and thus cause swelling, although the relative amounts of each which are present are not known. This process will alter the properties of the polymer, e.g., by decreasing the glass transition temperature. An flux of acetonitrile (dielectric constant of 37) would also be expected to cause an increase in the overall effective dielectric fonstant of the film compared to an unswelled film. Even if the dielectric constant of the swelled film was known, however, there would still remain the important question of heterogeneity. Local microenvironments could exist in the swollen, reduced film which exhibit dielectric and solvation behavior significantly different from those associated with averaged, bulk properties. Any oxidation of the LPVF film would, in addition to increasing the a ferricinium/counteranion concentration, also cause the amounts of sorbed solvent and supporting electrolyte to undoubtedly change. Important physical properties of the LPVF film are thus expected to be a function of its potential-controlled oxidation state.

With these caveats in mind, a qualitative picture of redox site aggregation in LPVF films can be given. In order to gain some understanding as to how various types of redox site associations could alter the equilibrium redox behavior of an LPVF film, the simplest cases of dimer formation are first considered. The simplest ionic dimer would be comprised of two Fc^+ ClO_4^- ion pairs and would be termed a quartet in Eisenberg's terminology[13]. Such an association for a redox system can be treated as oxidant dimerization, and the appropriate equilibrium expression is[14]:

$$E = E^{o'} + \frac{RT}{nF} [\ln \frac{2f}{1-f} - \ln(1 + (1 + 8fc_{Fc}K_{do})^{=})] \quad (4)$$

where f is fraction of sites oxidized, c_{Fc} is the total concentration of ferrocene sites, and K_{do} is the equilibrium constant for the dimerization of the oxidant Fc^+ (actually $Fc^+ \; ClO_4^-$). The lower right portion of Figure III.5.9 shows equation (4) plotted for several values of $c_{Fc}K_{do}$. It can be seen that oxidant dimerization shifts the formal potential to more negative values and causes a decrease in Nernst slope for $f \leq 0.5$.

Reductant dimerization is a second possibility for which a comparable expression to equation (4) yields the results shown in the upper left portion of Figure III.6.9 for various values of $c_{Fc}K_{dr}$, where K_{dr} is the equilibrium constant. Dimerization of two reductant sites, i.e., $2Fc^o \rightleftharpoons Fc_2^o$ will shift the formal potential to more positive values and will cause a decrease in Nernst slope for $f > 0.5$.

If the values of K_{do} and K_{dr} are both nonzero, the resultant Nernst plot will reflect the contributions of each according to the expression [14]:

$$E = E^{o'} + \frac{RT}{nF} \ln \frac{(1+8fc_{Fc}K_{do})^{1/2} - 1}{(1+8(1-f)c_{Fc}K_{dr})^{1/2} - 1} \qquad (5)$$

Results for the special case $K_{do} = K_{dr}$ are given in Figure III.6.10. No shift in the formal potential occurs and the decreased Nernst slope is symmetrical about f = 0.5. When $K_{do} \neq K_{dr}$, an $E^{o'}$ shift will occur and the Nernst plot again will be nonsymmetrical about f = 0.5. The extent to which these

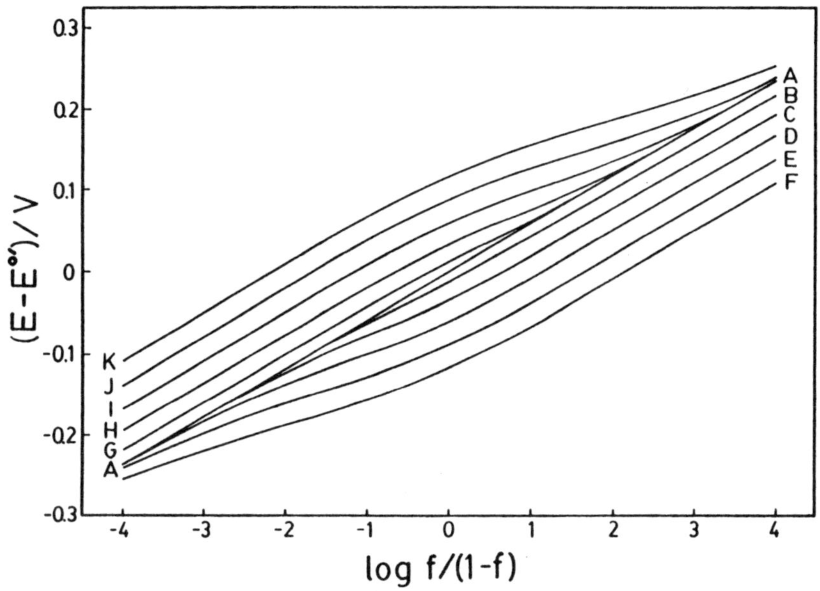

Figure III.6.9: Nernst plots showing the effect of oxidant or reductant dimerization. Curve A-$K_{do}=K_{dr}=0$. Curves B-F for oxidant dimerization with values of $c_{Fc}K_{do}$ equaling 1, 10, 10^2, 10^3 and 10^4, respectively. Curves G-K for reductant dimerization with values of $c_{Fc}K_{dr}$ equaling 1, 10, 10^2, 10^3 and 10^4, respectively. A one electron process is assumed in all cases.

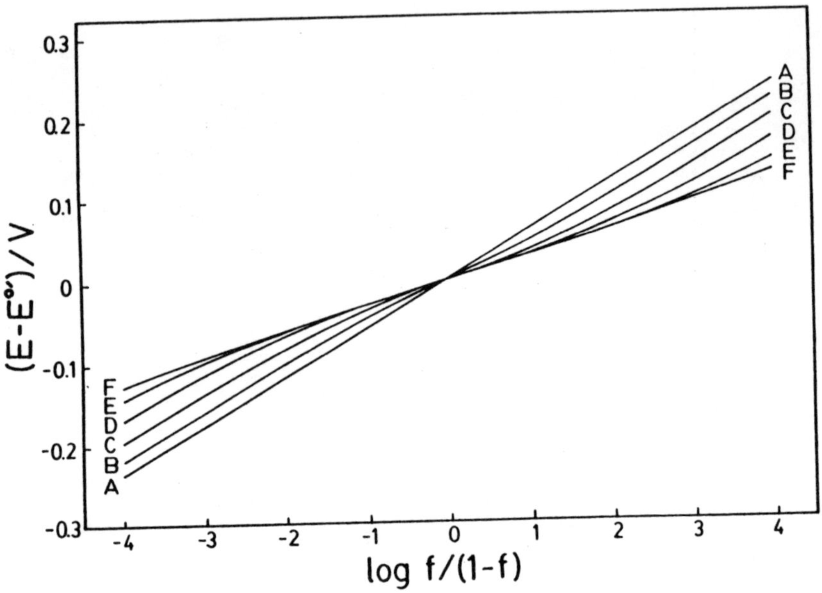

Figure III.6.10: Nernst plots reflecting committant oxidant and reductant dimerization. Values of $K=K_{do}=K_{dr}$ for curves A-F are 0, 1, 10, 10^2, 10^3 and 10^4, respectively.

deviations will occur will depend on the magnitudes of K_{do} and K_{dr}, and the resulting plot will more closely resemble that form of dimerization which corresponds to the larger of the two equilibrium constants.

In Region II, the Nernst slope is less than theoretical for $f \leq 0.5$, which indicates that oxidant dimerization may be occurring. Figure III.6.11 shows that equation (4) with $c_{Fc}K_{do}$ = 100 fits this data region quite well. We emphasize at this point that no quantitative significance is being attached to this particular value for two reasons. First, the shape of the nernst curve over this limited region is relatively insensitive to the magnitude of $c_{Fc}F_{do}$ (see Figure III.6.9 for $0 > \log(f/(1-f)) > -1.5$). The second reason is that the shift in formal potential, relative to the simple case, is not known (see below). Nevertheless, Figure III.6.11 does serve to show that a sub-Nernstian slope in Region II can be qualitatively viewed as dimerization of oxidized redox sites, i.e., formation of $Fc^{+}\ ClO_4^{-}$ ion pairs. Dimerization of the reduced sites Fc^{o}, on the other hand, cannot account for the reduced slope as can be concluded from inspection of Figure III.6.9. One other case that has been considered as a simple model for redox site aggregation is a dimer comprised of one oxidant and one reductant[14], e.g., $Fc^{o/+}\ ClO_4^{-}\ Fc^{+/o}$. However, this model incorrectly predicts a super-Nernstian slope about f = 0.5 and has, therefore, been eliminated from further consideration.

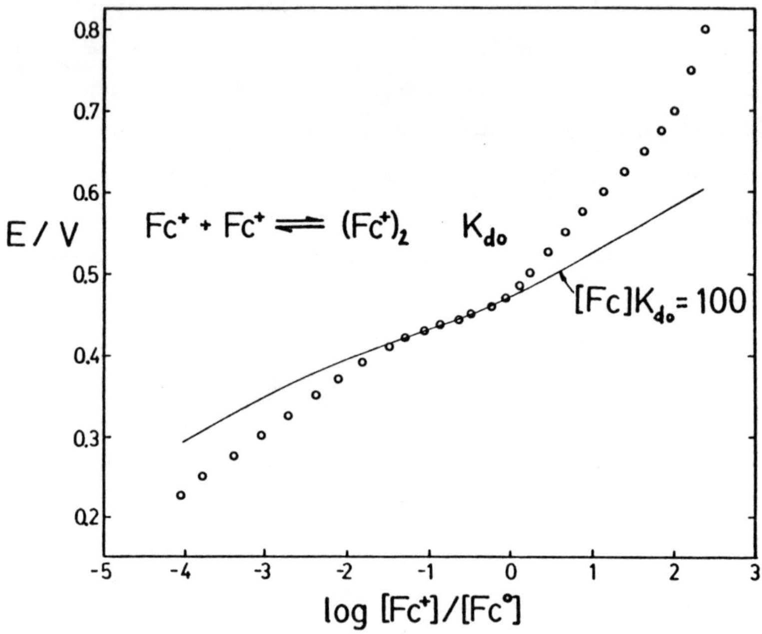

Figure III.6.11: Fitting of oxidant dimerization model to Region II in a 745 Å film of LPVF on glassy carbon. SSE: 0.10 m TBAP/acetonitrile.

The preceding discussion has enceavored to show that ion association is an expected phenomeon in LPVF films (as in other ionic polymers), and that such association is qualitatively consistent with predicted and observed redox behavior by using a simple dimerization model. A closer inspection reveals that oxidant dimerization (quartet formation) cannot, by itself, account for the available data. For example, minimum slopes of 20-25 mV were consistently obtained for LPVF Region II data, and, in fact, the true minimum slope is probably even less than this graphically determined value whose accuracy is limited by the width of the intervals between data points. Since dimerization alone could afford a minimum slope of only 29.6 mV for large values of k_{do}, the formation of higher order multiplets appears to be likely. The minimum attainable Nernst slope for a multiplet composed of n $Fc^{+} \cdot ClO_4^-$ ion pairs would be 59.2/n mV.

Another important consequence of oxidant dimerization mentioned above and shown in Figure 9 is the strong dependence of the shift in $E^{o'}$ on the magnitude of K_{do}. Quantitative assessment of this effect requires a knowledge of $E^{o'}$ for the Fc^+/Fc^o couple in the polymer film in the absence of complications. In solution phase systems, dilutions of the redox couple are typically performed in order to obtain this information. Unfortunately, in LPVF and other redox polymer films dilution experiments are usually unavoidably ambiguous because of uncertainties with regard to accompanying changes in polymer structure. Comparisons of $E^{o'}$ values for redox couples in dilute solution and inside polymer films do, however, suggest that they are similar[7]. Also, the $E^{o'}$ of the ferrocene/ferrocinium couple is considered to be reasonably insensitive to its solvent environment[15], although certainly not totally so[16]. Thus, $E^{o'}$ for Fc^+/Fc^o at infinite dilution in LPVF is expected to be similar to its corresponding solution value. In 0.10 M TBAP/acetonitrile the solution $E^{o'}$ was determined from CV to be +0.41 V vs Ag/AgCl(1.0M KCl). From coulometry, the measured value for LPVF was +0.44 V. Because these two values are close to each other, it can be concluded that a large $E^{o'}$ shift cannot be attributed to the dimerization or association phenomena occurring in the LPVF film. For example, if the simulation shown in Figure III.6.11 is assumed for the moment to be true, an $E^{o'}$ shift of -60 mV (see Figure III.5.9) would occur. This would require a "dilute $E^{o'}$" for Fc^+/Fc^o in LPVF of 0.50 V. Larger values of K_{do} or higher order association would demand even larger values for the $E^{o'}$ at infinite dilution. In view of the earlier arguments, such a situation seems unlikely.

However, if some form of reductant dimerization or association were also present, then significantly decreased Nernst slopes could result without large formal potential shifts, e.g., Figure III.6.10. Physically, this is not unreasonable. As the ionic content of LPVF increases, the tendency of nonionized regions to coalesce into their own domains could effectively be considered as association. Such a situation has been proposed as inherent features of ion multiplet and cluster formation in ionic polymers[9,13].

Finally, with regard to Region II, some hysteresis evidence is presented which is supportive of the existence of a heterogeneous microphase environment in LPVF.

For Region III of LPVF redox behavior, mechanical forces are proposed as the basis for the observed increase in slope to values of ca. 110 mV/decade. As a reduced LPVF film is progressively oxidized, this phenomenon is proposed to first appear at a potential near $E^{o'}$. In essence the model assumes that for further oxidation of the film beyond this point to occur, counterions which enter the film do so only with a concommittant 3-dimensional elastic expansion of the polymer network, which is electrostatically crosslinked. In other words, mechanical work must be done against the electrostactically crosslinked LPVF film in order to further oxidize Fc^o sites in Region III. Alternatively, one may view the overall process of $Fc^o \rightarrow Fc^+$ conversion in this region to require not only the expenditure of electrochemical work, but also the expenditure of mechanical work. These two work terms are then summed to account for the overall free energy changed as reflected in the change in phase potential of the polymer film. We refer to this model as a mechanical/electrochemical (MEC) model.

The route by which the mechanical work term manifests itself electrochemically is proposed to be electrostatic charging of the film. Assume that an LPVF film is at equilibrium at some potential in Region III and is in a state of elastic strain. The retractive force resulting from polymer chain extension must be opposed by an expansive force of equal magnitude. This expansive force arises from crowding of counterions into the film to maintain charge neutrality. The driving force for sustaining the expansive force appears to be electrostatic in nature with the film becoming positively charged. In effect, there will be a small number of Fc^+ sites which remain unneutralized by anions. Thus, for LPVF, a 110 mV slope means that an additional 50 mV of potential is required to achieve the same degree of oxidation beyond that for the simple one electron case. This additional 50 mV is the change in film phase potential required to sustain the internal film pressure against the retractive force described above.

Before proceeding with the details of the MEC model, some comments about its development are given here. The MEC model was initially conceived for PPVF films, which are described more fully in a later section. For the present discussion, the salient features of PPVF films are their super-Nernstian (ca. 130 mV) slopes obtained from coulometric titration measurements, the invariance of this slope and the $E^{o'}$ with respect to various solvent/supporting electrolyte systems, and their insolubility in all solvents tested. These and other results lead to a picture of PPVF as a highly crosslinked polymer which is poorly solvated by all solvents used. An important distinction between the PPVF films and LPVF Region III behavior is the nature of the crosslinking. It is covalent in the case of PPVF, and thus found to predominate at all fractions oxidized; whereas in LPVF it is electrostatic and therefore only important at higher fractions oxidized.) It seemed reasonable to assume that volumetric expansion of the PPVF network would be required in order to accommodate concentrations of perchlorate ranging as high as 4 M. Because of the highly crosslinked structure of PPVF, it also seemed reasonable to assume that expansion of the film would be opposed by elastic mechanical forces resulting from stretching of the polymer network chains. Some particularly graphic evidence in support of this mechanism are found in Figure III.4.14. The scanning electron micrographs reproduced there reveal that PPVF films can indeed rupture upon oxidation. Such events appear to result from internal mechanical stress which exceeds the fracture stress of the material.

In contrast to PPVF, there are no covalent crosslinks in LPVF to provide a network structure. However, the dimerization or aggregation of $Fc^{+} \cdot ClO_4^-$ ion pairs described earlier serves to electrostatically crosslink the linear chains as is generally described for other ionic polymers[9]. Thus an LPVF film will undergo conversion from a linear polymer into a network polymer as it becomes oxidized. Upon oxidation of 50% of the redox sites in LPVF, the polymer should be extensively crosslinked since the concentrations of Fc^+ sites and ClO_4^- counterions are on the order of 3 M. Further influx of counterions will require work to elastically expand the resulting network.

It is not understood why the MEC effect is manifested only for $f \geq 0.5$. Undoubtedly there are a number of competing processes which occur as a reduced LPVF film becomes progressively oxidized. A reduced film will contain a certain fractional volume of solvent. As a film becomes more oxidized, a process which gradually transforms LPVF into a network polymer, the extent of solvation should decrease as the effective molecular weight of the polymer increases. Also upon oxidation of Fc^o sites, the corresponding volume of incorporated counterions (and their attendant solvation sphere) must be compensated for by either an increase in the overall volume of the film or by a loss of solvent from the film or by some combination thereof. For values of $f \leq 0.5$, the incorporation of anions evidently occurs in a manner which results in polymer structures which remain essentially unstressed. Beyond this point the film expands elastically to accommodate the increased volume associated with additional counterion influx. The transition from Region II to Region III redox behavior may correspond to a situation in which essentially all solvent has been exlcuded from the film. A competition exists between solvation effects and electrostatic effects, with the latter prevailing up to the point where the film may be viewed as no longer solvated internally.

The following paragraphs outline a crude quantitative treatment of Region III redox behavior in terms of a simple stress-strain model. The development consists of three major parts. First, the pertinent equations for elastic stress-strain behavior are presented. Next, a relation is given which relates strain (E) in an expanded LPVF film to the variable f, which is the fraction of the redox sites which are oxidized. Finally, the mechanical component of electrochemical free energy is calculated as a function of f (and therefore, E) and compared to that expected on the basis of simple elastic stress-strain behavior.

As a model for the LPVF film which is strained due to electrochemically induced internal expansive forces, we shall use the case of a body which is strained elastically under equal triaxial tension. This is a special case of the more general situation referred to as elastic volumetric (or dilatational) strain[17,18]. The body is assumed to be isotropic, isothermal, and elastic for the stresses considered. Application of a hydrostatic stress, either tensile or compressive, will result in a nondistorted, volumetric change, ΔV, relative to the original volume, V_o. The three orthogonal principal stresses will be equal:

$$\sigma = \sigma_x = \sigma_y = \sigma_z \qquad (6)$$

as will the corresponding principal strains:

$$\varepsilon = \varepsilon_x = \varepsilon_y = \varepsilon_z \qquad (7)$$

For the remainder of this paper the symbols σ and ε will be used to denote the respective one-dimensional stress and strain components for the volumetric strain case just described.

The volume change is related to ε by the following equation [18]:

$$3\varepsilon + 3\varepsilon^2 + \varepsilon^3 = \Delta V/V_o \qquad (8)$$

For the maximum relative volume change considered in the present work, i.e., 10%, an error of only 3% is introduced by ignoring the quadratic and cubic strain terms. Therefore, the following approximation is used instead of equation (8):

$$\varepsilon = \Delta V/3V_o \qquad (9)$$

The relationship between stress and strain for hydrostatic volumetric strain is given by:

$$\sigma = E_y\varepsilon/(1-2\mu) \qquad (10)$$

where E_y is Young's modulus and μ is Poisson's ratio.

For an elastic deformation of a body, the available work is the Helmholtz free energy, F [19]:

$$F = \frac{1}{2}\sigma_{ik}\varepsilon_{ik} \qquad (11)$$

where σ_{ik} and ε_{ik} represent the stress tensor and strain tensor, respectively. For hydrostatic deformation of an isotropic body, only the three diagonal (or principal) components of each tensor are nonzero and, by using equations (6) and (7), the expression for free energy becomes:

$$F = \frac{3}{2}\sigma e \qquad (12)$$

Substituting for σ from equation (10):

$$F = 3E_y\varepsilon^2/2(1-2\ ) \qquad (13)$$

For one-dimensional elastic strain, free energy is the area under the stress-strain curve, i.e., $\frac{1}{2}\sigma\varepsilon$. For hydrostatic volumetric elastic strain, the free energy is simply three times that value, as shown in equation (12). Equation (13) shows that the free energy is quadratic in the strain component [19]. Taking the square root of both sides of equation (13) gives the final equation of interest:

$$F^{1/2} = [3E_y/2(1-2\mu)]^{1/2}\varepsilon \qquad (14)$$

Attention is now turned towards the relationship between ε and f for a LPVF film in Region III. To simplify calculations, it will be assumed that the film is unstressed for $f \leqslant 0.5$. For $0.5<f<1$, ε will be assumed to result solely from the relative volume increase associated with the uptake of perchlorate couterions. Therefore, as f changes from 0.5 to 1, ε will change from 0.0 to ε_{max}:

$$\varepsilon = 2(f-0.5)\varepsilon_{max} \qquad (15)$$

To calculate ε_{max} the approximation is made that the concentration of ClO_4^- in the film increases from 3 M to 6 M as f changes from 0.5 to 1. Using equation (9) with volumes replaced by molar volumes and appropriate expressions inserted for these, the result is:

$$\varepsilon_{max} = \frac{1}{2}(4\pi r^3 N_A/3)/(3/C_{Fc}) \quad (16)$$

where r is the radius of the perchlorate ion, N_A is Avogadro's number, and c_{Fc} is the total concentration of ferrocenic sites in the film. The factor 1/2 appears since elastic strain occurs only for $0.5<f<1$. Using a value for r of 2.4 Å [20] and c_{Fc} = 6 M, ε_{max} is calculated to be 0.035. With this value, equation (15) can now quantitatively relate elastic strain to the fraction oxidized of an LPVF film subject to the assumptions made in the derivation.

The third major part of this treatment is calculation of the mechanical component of the change in electrochemical free energy. To simplify matters, LPVF Region III (see Figure III.6.8) will be approximated by a line originating at the $E^{0'}$ and having a slope of 110 mV. This situation, along with the simple one electron Nernstian case, are shown in Figure III.6.12 in terms of $(E-E^{0'})$ vs f. The cross-hatched area of Figure III.6.12 is assumed to be proportional to the mechanical contribution of the electrochemical free energy change. The symbol $\Delta\tilde{G}_m$ is used to designate the

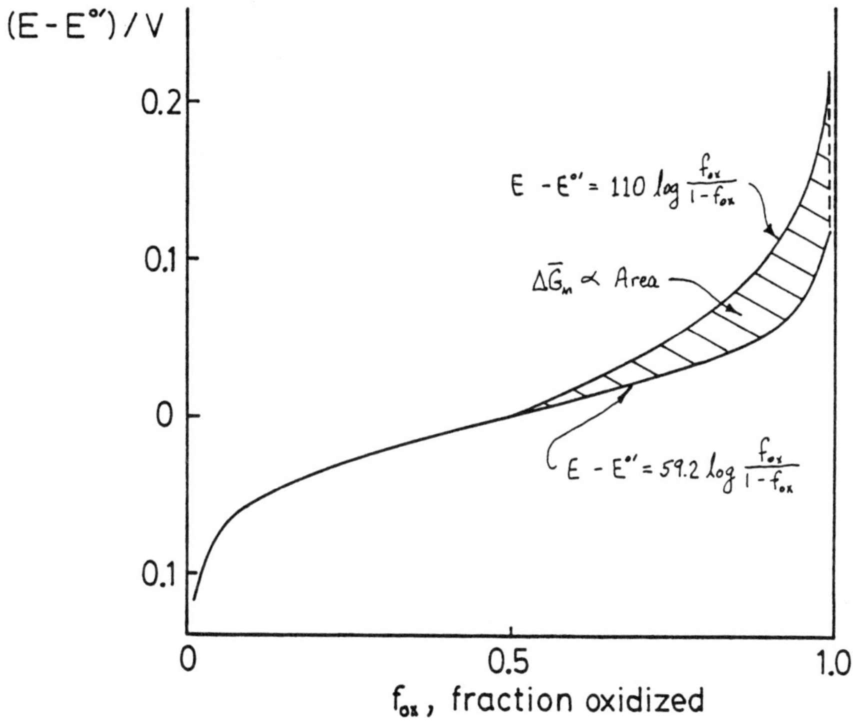

Figure III.6.12: Relationship between $(E-E^{0'})$ and f_{ox} the ideal one electron process compared to the observed behavior in Region III.

mechanoelectrochemical free energy per mole, and its value can be evaluated according to:

$$\Delta G_m = \int_{0.5}^{f} -nF(E_1-E_2)df \qquad (17)$$

Inserting the expressions for E_1 and E_2 gives:

$$\Delta G_m = (51\ mV)nF \int_{0.5}^{f} \log \frac{f}{1-f}\, df \qquad (18)$$

Solving the integral and making appropriate unit conversion results in:

$$\Delta G_m = (4.92x10^{10}\ ergs/mole)\ (f\log f+\log(1-f) - f\log(1-f)]_{0.5}^{f} \qquad (19)$$

In Figure III.6.13, ΔG_m and $(\Delta G_m)^{1/2}$ are graphed vs f (and ε by using equation (15)). For values of f approaching 0.9, $(\Delta G_m)^{1/2}$ vs ε can be approximated by a line as shown. Using equation (14) and assuming that μ=0.33 [21], the slope of this line can be used to calculate a value for E_γ of 1.2 x 10^{10} dyn/cm^2.

There are several reasons why this value of E_γ can only be considered as a rough estimate. (1) Region III redox behavior in the absence of mechanical effects is unknown. It is an oversimplification to use a 59 mV Nernst line since aggregation effects are also likely to be occurring in Region III. Although the extent of aggregation is unknown, its effect (see previous

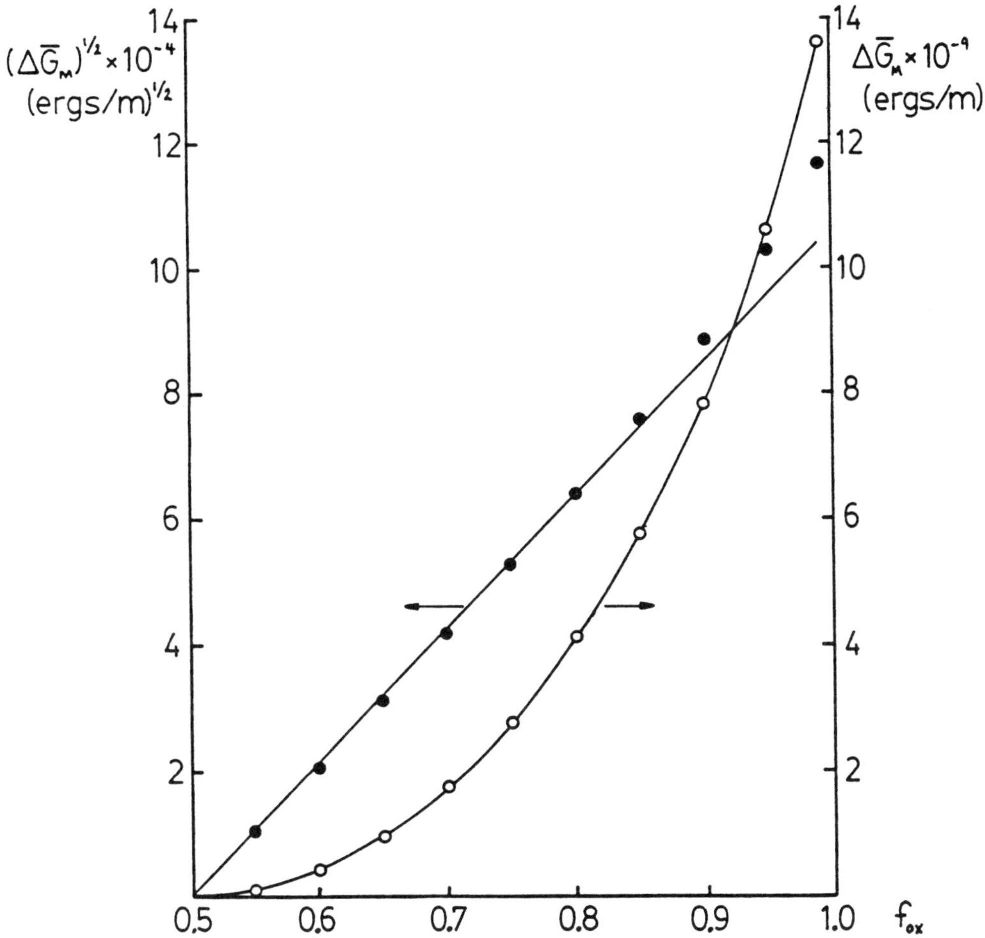

Figure III.6.13: $\Delta\bar{G}_M$ as a function of f_{ox} on the interval f_{ox}=D.5 to 1.0 (right ordinate). Left ordinate shows fit of $(\Delta\bar{G}_M)^{1/2}$ as a function of f_{ox} in this region.

section) will be to lessen the Nernst slope and increase the calculated value of E_y. (2) E_y is most likely dependent to some degree on the LPVF oxidation state. (3) The LPVF situation is not a true hydrostatic situation since one side of the film is constrained to some unknown degree by the electrode surface. (4) The value of Poisson's ratio for these films is not known and may be significantly different than the assumed value of 0.33.

With these foregoing limitations in mind, we feel that it is reasonable to conclude that the major results obtained from the MEC model provide good support for its validity. First of all, the manner in which free energy manifests itself as a function of f is similar to that expected from simple elastic stress-strain behavior; see Figure III.6.13. Secondly, a value for E_y of 1.2×10^{10} dyn/cm^2 is comparable to E_y values reported for glassy amorphous polymers [22]. As an order of magnitude estimate, it is entirely reasonable. In all likelihood, the actual value of E_y is higher than that calculated here insce LPVF in Region III is highly ionic and extensively crosslinked while amorphous glassy polymers are not. This discrepancy is undoubtedly a result of one or more of the error sources described above.

III.6.48

Region I: Ion Exchange Membrane Model

The details of the thermodynamic behavior of LPVF films in Region I are more poorly understood than those of Regions II and III. As can be seen from Figure III.6.14, not only is a super-Nernstian slope observed at log $[Fc^+]/[Fc^o] < -1.5$, but also this slope decreases with increased concentration of electrolyte in the solution phase. These data, which have been corrected for change in junction potential with change in [TBAP], strongly suggest that the potential-concentration relationship in Region I (<3% Fc^+) is controlled both by the change in electrochemical free energy of the ferrocene and ferricinium sites and the change free energy of transfer of electrolyte ions between the partially oxidized film and solvent. An ion exchange phenomenon in which the fixed ferricinium site concentration may be modulated by the phase potential of the LPVF film is consistant with these observations.

If it is assumed that there are no significant free energy contributions aside from the electrochemical interconversion of ferrocene/ferricinium sites, one would anticipate a 59 mV/decade slope for Figure III.6.14. The "excess slope" which observed is attributed to the membrane phenomenon alluded to above. If one calculates the free electrolyte ion concentration in the solution phase by considering the tendency for TBAP to ion pair in

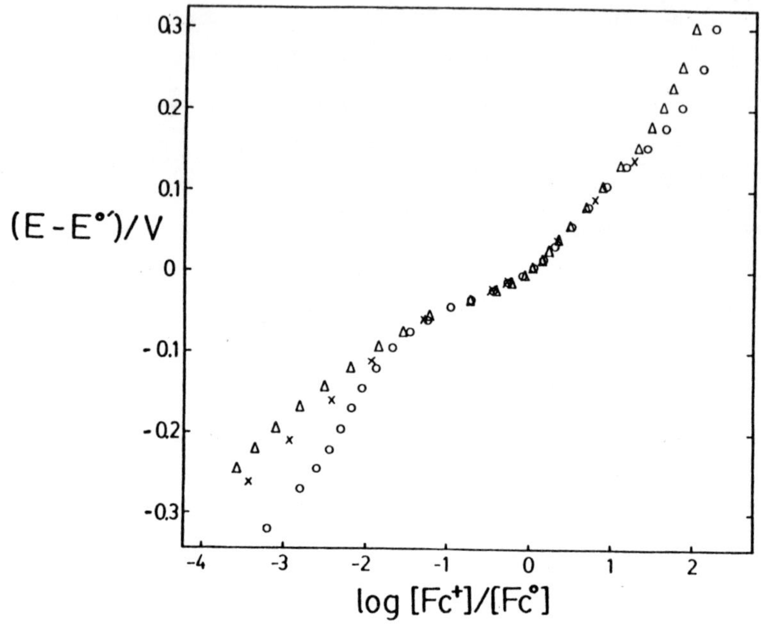

Figure III.6.14: Nernst plots for a 490 Å film of LPVF supported on glassy carbon. In all cases the solvent-supporting electrolyte system was TBAP/acetonitrile. Triangles are for data obtained at [TBAP] = 0.10 M, crosses at [TBAP] = 0.10 M and circles at [TBAP] = 0.010 M. Respective slopes in Region I are 75, 100 and 165 mV/decade.

acetonitrile[23], then one finds the "excess slope" to vary as the log of the free ion concentration. Furthermore, this relationship has a near Nernstian slope of ca 55 mV/decade. What is puzzling, however, is that the contribution of the ion exchange phenomenon to the overall potential drop from the supporting (glassy carbon) electrode to the electrolyte solution opposes the Faradaic contribution. This relationship dictates that there be excess positive charge in or on the film. One might have expected that, since cationic sites are being introduced by the Faradaic conversion of ferrocene to ferricinium ions, the membrane so-produced would respond to the perchlorate ion. Thus, the film would show an increasingly negative deviation from the Faradaic response line with increasing free ion concentration (c.f Figure III.6.14).

That the membrane responds to tertra-n-butylammonium ion (TBA^+) leads us to conclude that a double layer must form at the polymer-solution interface at low extents of conversion of reduced LPVF to the oxidized form. The implications of this are that the majority of the cationic sites formed in Region I are confined to the outermost (solution side) region of the film, and that perchlorate then specifically and strongly adsorbs to these sites in a manner such that the negative charge on the perchlorate is not efficiently screened by the cationic sites within the polymer.

Completion of the double layer structure follows when TBA^+ then adsorbs to these unscreened anionic sites.

This structure is then viewed to collapse as the extent of LPVF oxidation exceeds ca 3%, at which point there is a rapid influx of solvent and supporting electrolyte. Thus, the structural reorganization which accompanies the transition from Region I to Region II is rather extensive. Furthermore, the oxidation process would appear in Region I to occur from the polymer-solution interface within the context of this model.

References

1. Peerce, P. J. and Bard, A. J., J. Electroanal. Chem, 108, 121 (1980).

2. Flanagan, J. B., Margel, S., Bard, A. J. and Anson, F. E., J. Amer. Chem. Soc., 100, 4248 (1978).

3. Merz, A. and Bard, A. J., J. Amer. Chem. Soc. 100, 3222 (1978).

4. Daum, P. and Murray, R. W., J. Phys. Chem., 85, 389 (1981).

5. Nowak, R. J., Schultz, F., Umana, M. and Murray, R. W., Anal. Chem., 52, 315 (1980).

6. Brady, J. M., University of Minnesota, unpublished results.

7. Schroeder, A. H., Kaufman, F. B., Patel, V. and Engler, E. M., J. Electroanal. Chem., 113, 193 (1980).

8. Peerce, P. A., and Bard, A. J., J. Electroanal. Chem., 114, 89 (1980).

9. Eisenberg, A. and King, M., Ion-Containing Polymers, Academic Press, New York, 1977.

10. MacKnight, W. J. and Earnest, T. R., J. Polym. Sci., Macromol. Rev., 16, 41 (1981).

11. Holliday, L., Holliday, L. (ed.), Ionic Polymers, John Wiley and Sons, New York, 1975, p.1.

12. Van Krevelen, D. W., Properties of Polymers-Correlations With Chemical Structure, Elsevier, Amsterdam, 1972, pp. 213, 215.

13. Eisenberg, A., Macromolecules, 3, 147 (1970).

14. Clark, W. M., Oxidation-Reduction Potentials of Organic Systems, Krieger, Huntington, NY, 1972, Chapt. 6.

15. Laitinen, H. A. and Harris, W. E., Chemical Analysis, 2nd Ed., McGraw-Hill, New York, 1975, p. 59.

16. Sahami, S. and Weaver, M. J., J. Solution Chem., 10, 199 (1981).

17. Polakowski, N. H. and Ripling, E. J., Strength and Structure of Engineering Materials, Prentice-Hall, Englewood Cliffs, NJ, 1966, Chapter 5.

18. Shanley, Strength of Materials, McGraw-Hill, New York, 1957, Chapter 9.

19. Landan, L. D., and Lifshitz, E. M., Theory of Elasticity, 2nd English Ed., Addison-Wesley, Reading, MA, 1970, pp. 10-12.

20. Jenkins, H.D.B. and Thakur, K. P., J. Chem. Educ., 56, 576 (1979).

21. Van Krevelen, D. W., ref. 12, p. 157.

22. Van Krevelen, D. W., ibid., p. 150.

23. Pemberton, J. E., McIntire, G. L., Blount, H. N. and Evans, J. F., J. Phys. Chem., 83, 2696 (1979).

III.7 Transport Kinetics in PPVF Film Electrodes

Break-In Behavior - The importance of the first scan recroded in cyclic voltammetric experiments with electroactive polymer modified electrodes was emphasized earlier. The break-in behavior may be related to properties of the polymer film (such as polymer film density and extent of crosslinking), as evidenced by the overpotential required to drive the electrochemical conversion of the electroactive polymer from one redox form to the other during the first few cycles [1].

In order to compare break-in behavior of PPVF films made in the second generation plasma reactor to those made in the prototype plasma reactor, the first scan cyclic voltammograms were recorded at 50 $mV\text{-}s^{-1}$. Very little break-in behavior was observed for PPVF films made in the new reactor and examined in 0.1 M $TBAP/CH_3CN$. This break-in was manifested in a small overpotential required in the first anodic sweep peak potential relative to the anodic peak potentials in subsequent scans. For the thinnest film examined (195 Å, $\Gamma_T = 8.63 \times 10^{-9}$ $moles/cm^2$) this overpotential was only 20 mV. For a 1394 Å film ($\Gamma_T = 6.25 \times 10^{-8}$ $moles\text{-}cm^2$) this overpotential was ca 100 mV. Greater than 95% of the ferrocene sites which are converted to ferricenium in subsequent scans on this cyclic voltammetric time scale underwent redox in the first recroded cyclic voltammogram. In sharp contrast to the above, the thickest PPVF film examined (4557 Å, $\Gamma = 1.84 \times 10^{-7}$ moles-cm^{-2}), no anodic wave was observed on the first scan between the potential limits used (0 → 0.9 volt vs. Ag/AgCl). Only 0.3% of the ferrocene centers were converted to ferricenium in the first scan with respect to the steady state fractional conversion on this cyclic voltammetric time scale (91% of the coulometrically determined coverage). A total of seven scans was required to reach a steady state cyclic voltammogram. Clearly, the transport properties of this film were changed as a consequence of electrochemical cycling. The scanning electron micrographs shown in Figure III.4.14 indicate that gross morphological changes occured, presumably due to this break-in. The thinner films (1400 Å or less) however showed no evidence for gross morphological changes, as determined by SEM or by scattering of the laser beam of the ellipsometer. It is not clear if the break-in observed and the rupture of the polymer film for this 4557 Å film was a consequence of the greater thickness of the film or if this film was more highly crosslinked. The amount of material extracted from this film was less than the other film for shorter deposition times and there was some uncertainty in steady state plasma pressure due to drift in the pressure gauge. Pressures lower than 6 mtorr would convert more of the as-deposited polymer to an insoluble polymer, possibly with a higher crossllink density. Therefore, a higher crosslink density might account for this pronounced break-in.

Comparing the break-in behavior of PPVF films made in the new reactor to those of equivalent coverage made in the prototype reactor, important diffrences are noted. The break-in behavior of a PPVF film made in the prototype reactor ($\Gamma = 2.7 \times 10^{-8}$ moles-cm^{-2}) is shown by the first scan which was very distorted. For a film made in the new reactor of comparable coverage (2.02×10^{-8} mole-cm^{-2}, d = 455 Å), the break-in behavior consisted of only 50 mV overpotential in the first recorded cyclic voltammogram. This implies that the film made in the new reactor was less crosslinked than that made in the old reactor. In both cases, the net RF power used was nearly the same (5 watts in the old reactor, 7 ±2 watts in the new reactor), but the volumes, and hence the power densities were quite different (20.4 watts/liter in the prototype reactor, 3.9 ±1/liter in the new reactor). The use of higher power density could account for higher degrees of crosslinking for films deposited in the prototype reactor.

The break-in behavior for films of the same thickness (made in the new reactor) was dependent on the solvent supporting electrolyte. 750 Å thick PPVF films showed little break-in (~50 mV at 50 mV-s^{-1}) in acetonitrile or methylene chloride using TBAP as the electrolyte. In contrast, a 750 Å film in acetonirile showed a much more pronounced break-in with TEApTs as the electrolyte under the same conditions. 68% of the ferrocene sites were converted to ferricenium in the first scan relative to subsequent scans. By comparison, Schroeder *et al.* [2] compared the effect of TEAP and TEApTs on the chronoamperometric response of 4000 Å thick phenoxy-tetrathiofulvalene films in acetonitrile. With TEAP, the film underwent rapid oxidation, but with TEApTs, the oxidation of the film was *completely inhibited*. Schroeder's observations underscore the importance of the counterion and its size relative to the free volume in the polymer film, and its effect on the redox behavior of electroactive polymer films. Although the drastic difference in electrochemical behavior between TBAP and TEApTs was not observed with PPVF films, lower concentrations of total ferrocene content in the PPVF film were found when TEApTs was the electrolyte (see previous section). These lower calculated concentrations are not due to variance in the film density, since the concentration calculated with TBAP as the electrolyte was reproducible (within 1.5%) for triplicate determinations (2 from the same deposition, 1 from another deposition, 15 minute deposition time). This suggests that only a certain volume of counterion may be accommodated in the PPVF films, and because of this, some of the ferrocene sites are electroinactive in the polymer film. Furthermore, these inactive sites do not participate in the determination of the thermodynamic behavior of PPVF films, since the same charge - potential relationship was found in the case of TEApTs and TBAP in acetonitrile. Although the PPVF films showed a pronounced break-in with TEApTs, no appreciable changes in the polymer film morphology were observed by laser light scattering and ellipsometry.

A pronounced break-in was also observed for a 779 Å PPVF film in $LiClO_4/H_2O$. Only 3% of the ferrocene sites were converted to ferricenium relative to steady state conversion at 50 $mV\text{-}s^{-1}$. A total of 9 scans were required to reach steady state, with the fractional amount converted gradually increasing with each scan. Water is a poor solvent for PPVF, and this break-in may reflect that very littel solvent is initially present in the film and that very little swelling occurs with water. The density (concentration of ferrocene) was also lower in the case of water. The lower concentration of ferrocene found might be a consequence of a lower free volume in the film (relative to that in acetonitrile) and hence less counterion can be accommodated. No appreciable changes in PPVF film morphology were observed in the case of water after electrochemical cycling.

It was mentioned previously that little break-in was observed in the case of a PPVF film in $TBAP/MeCl_2$. Based on solubilities of PVF in $MeCl_2$, it was expected that this solvent would be the best solvent of those studied here for PPVF, and hence the break-in should be minimized. However, the concentration of ferrocene measured in the solvent was lower than those measured in other solvents for films made in the same deposition ($[FeCp_2]$ in $MeCl_2$ for 765 A film = 3.6 M). If the film was swollen by $MeCl_2$, this would not be expected. After characterization in $MeCl_2$, this PPVF film was examined in TBAP/CH_3CN, and the surface coverage was found to be 8% higher than that found in $MeCl_2$. Recalculating the concentration of ferrocene in the film gave a value of $[FeCp_2]$ = 3.86 M, which is still significantly lower than that expected. The film thickness measured after characterization in $MeCl_2$ had decreased by 28 Å in the electrolyzed area. (Films studied in other SSE did not undergo any appreciable thickness change or change in nf after electrochemical characterization.) Assuming this decrease in thickness was due to loss of electroactive ferrocene, the $[FeCp_2]$ was adjusted for this loss and a value of $[feCp_2]$ = 4.04 M was calculated, which is still low compared to that expected (4.4 ± 0.06). The refractive index of the film was 1.68 after electrochemical characterization in CH_2Cl_2, as opposed to 1.70 prior to characterization. The lower refractive index suggests the film was less dense as a consequence of electrochemical cycling, which may account for the adjusted $[FeCp_2]$ still being low.

The decrease in thickness and refractive index implies that part of the PPVF film dissolved in $MeCl_2$. This material which dissolved could be lower molecular weight ferrocene polymer which leached out as a consequence of electrochemically induced swelling. However, this dissolution of part of the PPVF film is in contradiction with the observation that less of the film is electroactive in CH_2Cl_2 than in CH_3CN, suggesting that CH_2Cl_2 is a poorer swelling solvent. It is known that CH_2Cl_2 is a good solvent for neutral forms of PVF, but a poor solvent for the oxidized form of PVF. This redox state dependent solubility of PVF is the basis for fabrication of PVF modified electrodes by electrodeposition [3]. In light of this, it is conceivable that, as the PPVF film is oxidized, CH_2Cl_2 becomes a poorer solvent, and the swelling of PPVF by CH_2Cl_2 decreases at more anodic potentials, possibly decreasing the polymer free volume. This in turn hinders the influx of counterions and does not allow the film to become totally oxidized, causing some of the sites to be electroinactive and not participate in the overall electrochemical behavior of the film. The above potential dependent swelling may be an important variable to be considered in choosing solvents for examination of the electrochemical behavior of electroactive polymer modified electrode.

Charge-Transport Kinetics

A series of PPVF films in various solvent - supporting electrolytes was qualitatively evaluated as to their transient behavior in a cyclic voltammetric experiment. Only the thinnest film (195 Å) examined in TBAP/acetonitrile adhered strictly to a peak current linearly dependent on scan rate in the sweep rate range between 10 and 100 mV-s^{-1}. For all other films in all SSE, the peak current sweep rate dependence was intermediate between $i_p \propto \nu$ and $i_p \propto \nu^{1/2}$.

The ratio of anodic/cathodic peak currents was dependent on the solvent supporting electrolyte. For the thinnest filml in TBAP/CH_3CN $i_{p_a}/i_{p_c} = 1$ at sweep rates between 10 and 100 mV-s^{-1}. For thicker films in this SSE, i_{p_a}/i_{p_c} was greater than 1 (ca 1.09), and this value was relatively independent of film thickness. For films of equivalent thickness in various SSE systems, i_{p_a}/i_{p_c} ratio was as given in Table III.7.1.

Table III.7.1

d(Å)	SSE	i_{p_a}/i_{p_c} at 50 mV-s^{-1}
749 Å	0.1 M TBAP/CH_3CN	1.09
779 Å	0.1 M TEApTs/H_2O	1.07
762 Å	0.1 M TEApTs/CH_3CN	0.76
765 Å	0.1 M TBAP/CH_2Cl_2	0.88

Although this ratio is not very informative, it suggests that the relative rates of PPVF film oxidation an reduction depend on the solvent - supporting electrolyte, and are most probably related to the degree of swelling of the film by the solvent (which depends on its oxidation state) and the nature of the counterion.

Somewhat more informative is the peak potential separation as a function of sweep rate for films of different thickness and in different solvent supporting electrolytes. Figure III.7.1 shows the variation in peak potential as a function of sweep rate for films of different thickness in 0.1 M TBAP/CH_3CN, and Figure III.7.2 shows this for films of equivalent thickness in various solvent - supporting electrolytes.

Figure III.7.1 shows that the peak separation in all cases for various thickness films increases as a function of sweep rate, and that even for the thinnest film at the slowest sweep rate (10 mV-s^{-1}), the peak separation is not zero. This implies that the polymer film is not at equilibrium with the electrode, since according to theory ΔE_p should be zero [4]. In accordance with this, at 10 mV-s^{-1} only 93% of the total number of ferrocene sites (determined coulometrically) are converted to ferricenium for the 195 Å film. In the other extreme, at 100 mV-s^{-1} for the 1390 Å film, 82% of the film is converted on this time scale.

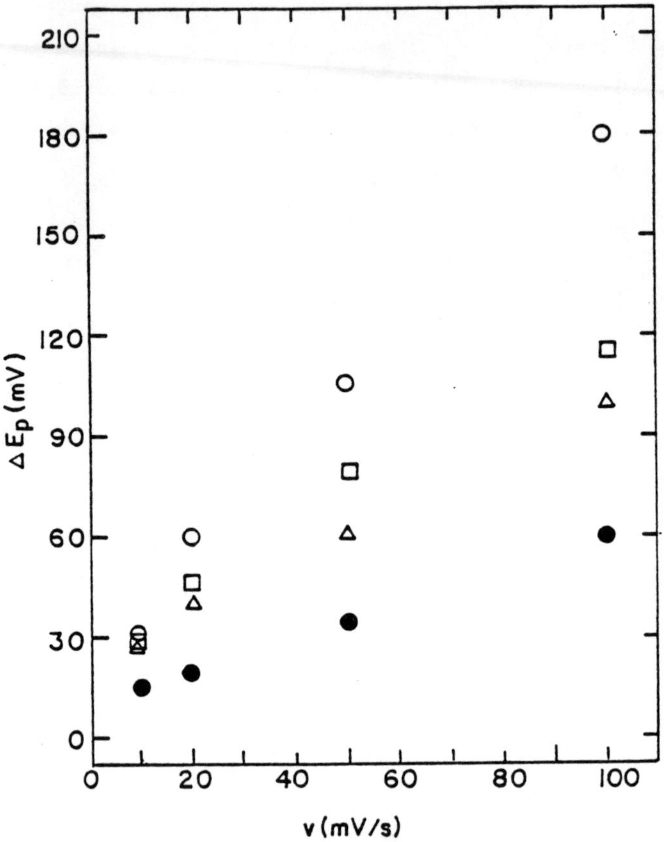

Figure III.7.1: Peak separation (ΔE_p) as a function of sweep rate (v) for PPVF films on glassy carbon of various thicknesses in 0.1 M TBAP/CH_3CN.

● = 195 Å PPVF film

△ = 473 Å PPVF film

□ = 749 Å PPVF film

○ = 1390 Å PPVF film

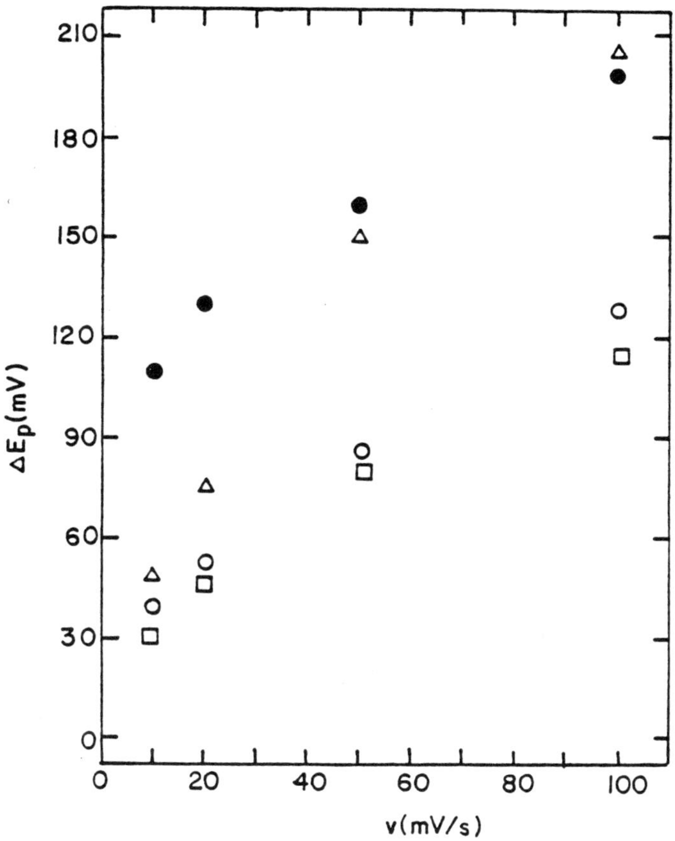

Figure III.7.2: Peak separation as a function of sweep rate for PPVF films of equivalent thickness in different solvent - supporting electrolytes.

□ = 749 Å PPVF film in 0.1 M TBAP/CH_3CN

○ = 762 Å PPVF film in 0.1 M TEApTs/CH_3CN

△ = 765 Å PPVF film in 0.1 M TBAP/CH_2Cl_2

● = 779 Å PPVF film in 0.1 M $LiClO_4/H_2O$

The full width at half maximum was relatively independent of film thickness at low sweep rate, but was broadened at higher sweep rates. In all cases, the full width at half maximum was greater than 120 mV. However, the FWHM was not equal for the cathodic and anodic peaks, the cathodic FWHM generally being 20 mV wider. This again suggests that on the time scales examined here, the PPVF film was not in equilibrium and charge transport limitations are implicated even for the thinnest films.

In Figure III.7.2 it is seen that the charge transport rate as evidenced by the peak separation strongly depends on the SSE used. The redox conversion appears fastest for TBAP/CH_3CN and is somewhat slower for TEApTs/CH_3CN. This suggests the counterion plyas an important role in the transport rate, which is to be expected if the transport rate is limited by the rate of diffusion of counterion through the PPVF film. That TEApTs causes slower redox conversion is also reasonable considering the break-in behavior observed with TEApTs which implies that the PPVF film undergoes a "rearrangement" in order to accommodate this electrolyte.

The transport rates are also affected by the solvent employed. Comparing the cases for water and methylene chloride to that in acetonitrile, all of which have ClO_4^- as the counterion, the transport rates are significntly slower for water and methylene chloride than for acetonitrile. Note however that the peak separation dependence on sweep rate is quite different in these respective solvents. In the case of water, there is a large peak separation (110 mV) at the lowest sweep rate examined (10 $mV\text{-}s^{-1}$) and the peak separation increases to 200 mV at 100 $mV\text{-}s^{-1}$. In the case of methylene chloride, the peak s eparation is signficantly smaller at low sweep rates (50 mV at 10 $mV\text{-}s^{-1}$) but increases more rapidly and is 210 mV at 100 $mV\text{-}s^{-1}$. The different dependence of peak potential separation on sweep rate in these solvents suggests different processes are responsible for the charge transport limitations in these solvents. In the case of water, which is an extremely poor solvent for the film (vinylferrocene and ferrocene monomers are not soluble in water), little or no swelling of the film might be implicated which may reduce the counterion transport rate. In the case of methylene chloride, electron transport might be the rate limiting process. Cyclic voltammetry of hydroxymethylferrocene in TBAP/CH_2Cl_2 shows irreversible behavior even at 10 $mV\text{-}s^{-1}$ (ΔE_p = 77 mV for ca 0.2 mM HMF). In all the other SSE, HMF showed reversible behavior.

The cyclic voltammetric peak shapes were also greatly affected by the SSE. For example, at 10 $mV\text{-}s^{-1}$, the anodic and cathodic FWHM in the case of $LiClO_4/H_2O$ was 110 mV and 155 mV respectively, and in the case of TEApTs/CH_3CN, the anodic and cathodic FWHMs were 190 mV and 160 mV respectively. Clearly, since the anodic and cathodic FWHM are not equal, this difference in FWHM is a kinetic effect, and not a consequence of thermodynamics. That the FWHMs might be broadened due to thermodynamics is not ruled out, but the anodic and cathodic FWHM should be equal to each other at equilibrium. Furthermore, since the FWHMs are different in different SSE, and if these FWHMs were purely a thermodynamic consequence, this would be reflected in the charge - potential relationship. In the preceedings section was shown that the thermodynamic behavior of the films was largely independent of SSE.

The observation that the anodic and cathodic peak currents and FWHMs are not equal for a given PPVF film electrode in a given solvent electrolyte implies that the kinetics of oxidation and reduction are not equal. This is not unreasonable, since the nature of the polymer film is different in its oxidized and reduced states. In the reduced state there is most likely very little supporting electrolyte, and probably solvent in the film, while in the oxidized state the concentration of counterion must equal that of the ferricenium present, which is $\sim$4.4 M, and the film is mot probably swollen. Further evidence for difference in the kinetics of oxidation and reduction is given by the chronoamperometric responses of these films. Figure III.7.3 shows the chronoamperometric transients for a 749 Å PPVF film in 0.1 M $TBAP/CH_3CN$. The potential was stepped from 0.0 to 0.9 volts, held for 2 minutes at 0.9 V, and then returned to 0.0 volts. A current maximum is seen in the anodic transient. The necessary implications is that the charge transport rate is initially slow and that as the film becomes partially oxidized, the charge transport rate increases.

However, in the cathodic peak, no maximum is seen. This demonstrates that the charge transport rate is different for the oxidation and reduction in PPVF films. This histeresis in anodic and cathodic chronoamperometric transients is a general effect for films of all thicknesses in all supporting electrolytes examined.

As recently suggested by Chambers [5], chronocoulometry is a facile method compared to chronoamperometry to evaluate effective diffusion coefficients for charge transport through polymer modifed electrodes. He derived an equation based on relations for coulometry on thin layers in solution, which was applicable to electroactive polymer modified electrodes. The equation derived by Chambers is given below.

$$\frac{Q}{Q_T} = 1 - \frac{8}{\pi^2} \exp(-\pi^2 \xi) \qquad (1)$$

where Q_T is the total charge consumed for exhaustive electrolysis, $\xi = t/\tau, \tau = (2d)^2/D$, where d is the polymer film thickness, D is

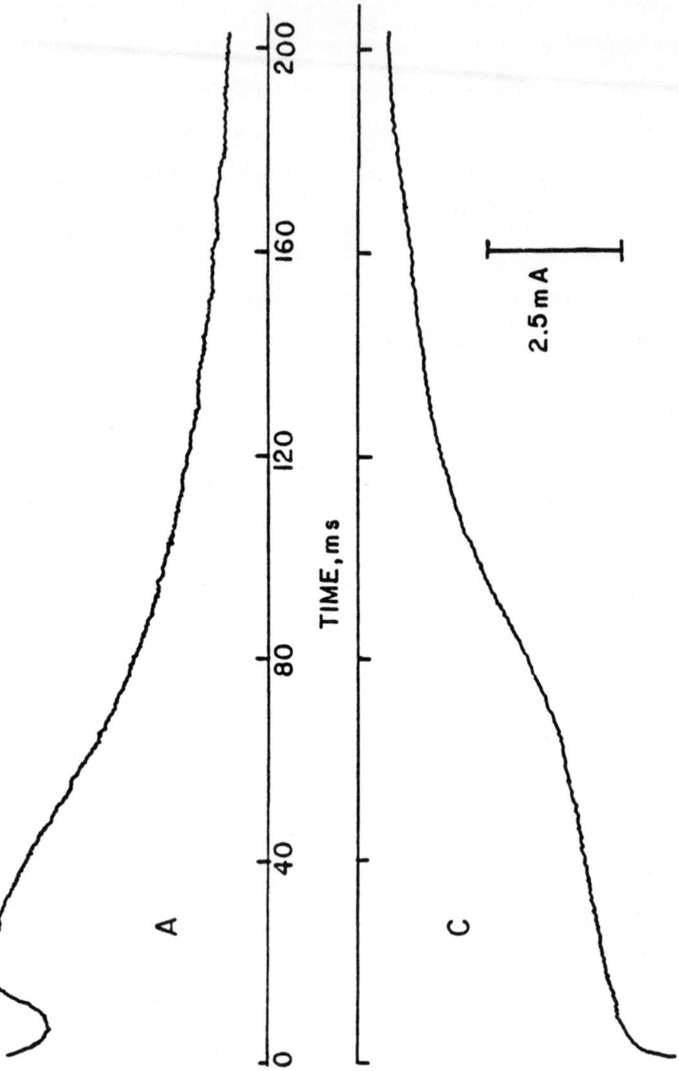

Figure III.7.3: Chronoamperometric response for a 749 Å PPVF film in 0.1 TBAP/CH_3CN.

A = Anodic pulse 0.00 to 0.09 volts

C = Cathodic pulse 0.09 to 0.0 volts

the charge transport diffusion coefficient (D_{eff}) and t is the time. The equation above is an approximation, and for better than 2% accuracy, should be used for $Q/Q_T > 0.33$. Chambers suggested the simplest way to use the above equation is to measure t at $Q = Q_T/2$. Here (at $t_{0.5}$), $\xi = \pi/8^2$ and

$$\tau = \frac{8^2}{\pi} \quad t_{0.5} \qquad (2)$$

Equation (2) was used here to measure the relative charge transport rate (D_{eff}) for PPVF films as a function of film thickness and SSE. The results are given below in Table III.7.2.

Table III.7.2

d(Å)	SSE	t(0.5)(s)[a]	D_{eff} (cm^2/s)
195	0.1 M TBAP/CH_3CN	0.022	3.39×10^{-11}
473	0.1 M TBAP/CH_3CN	0.08	5.49×10^{-11}
749	0.1 M TBAP/CH_3CN	0.09	1.22×10^{-10}
1390	0.1 M TBAP/CH_3CN	0.185	2.05×10^{-10}
762	0.1 M TEApTs/CH_3CN	0.42	2.71×10^{-11}
779	0.1 M $LiClO_4/H_2O$	0.193	6.17×10^{-11}
765	0.1 M TBAP/CH_2Cl_2	0.250	4.59×10^{-11}

(a) for anodic pulse 0.0 to 0.9 V.

For the results as a function of thickness, the agreement between calculated D_{eff} is very poor. Two factors may account for this. One is that there might be subtle changes in the PPVF film properties (e.g. crosslink density), which cause the diffusion coefficient to be different in PPVF films. That this might be true is evidenced by the $t_{0.5}$ values for the 473 Å and 749 Å films which differ only slightly (0.08 and 0.09 seconds), although the thickness varies by a factor of 1.6. Secondly, the model may not apply. It assumes that the polymer film does not change in thickness as a function of potential and that the diffusion coefficient (charge transport rate) is not a function of the oxidation state of the polymer film. The chronoamperometric transients show the latter is not true, and evidence was presented that the polymer films do swell. Hence, both the model and subtle differences in the PPVF films may contribute to the variation in the calculated diffusion coefficients. Nevertheless, comparisons of these calculated D_{eff} for PPVF films of equivalent thickness in different SSE still give a relative measure of the transport rate of these films in these solvents. Since these <u>ca</u> 750 Å films were all made in the same deposition, the ambiguity of possible subtle differences in the PPVF films is removed.

One rather surprizing finding is that the PPVF film in TEApTs/CH_3CN had the lowest values for D_{eff}. Based on the peak separation as a function of sweep rate (Figure III.7.2), the expectation was that the transport rate should be similar to that found for PPVF in TBAP/CH_3CN. However, the peak shapes were very different in both cases, having significantly broader peak widths in the case of TEApTs/CH_3CN. This is further evidence that the peak widths in TEApTs were a consequence of charge transport kinetics. For the case of water and methylene chloride, the peak widths were similar to those for the case of TBAP/CH_3CN, and hence the diffusion coefficient (charge transport rate) calculated is in the proper direction as that which would be inferred from the peak potential separation.

Substrate Transport Through PPVF Films

In order to evaluate the permeability of PPVF films, the oxidation of DPA was examined at PPVF film covered electrodes. DPA was chosen as the substrate for two reasons: (1) it has an $E^{o'}$ for its oxidation well removed from the $E^{o'}$ of PPVF (+1.2 V *vs.* Ag/AgCl); (2) comparisons could be made to the oxidation of DPA at PVF film covered electrodes [3].

Figure III.7.4 shows the oxidation of DPA at a bare Pt electrode and PPVF film covered electrode, with PPVF film thicknesses of 172 Å and 400 Å. Even with the thinnest film, the oxidation of DPA is greatly inhibited by the PPVF film (80%

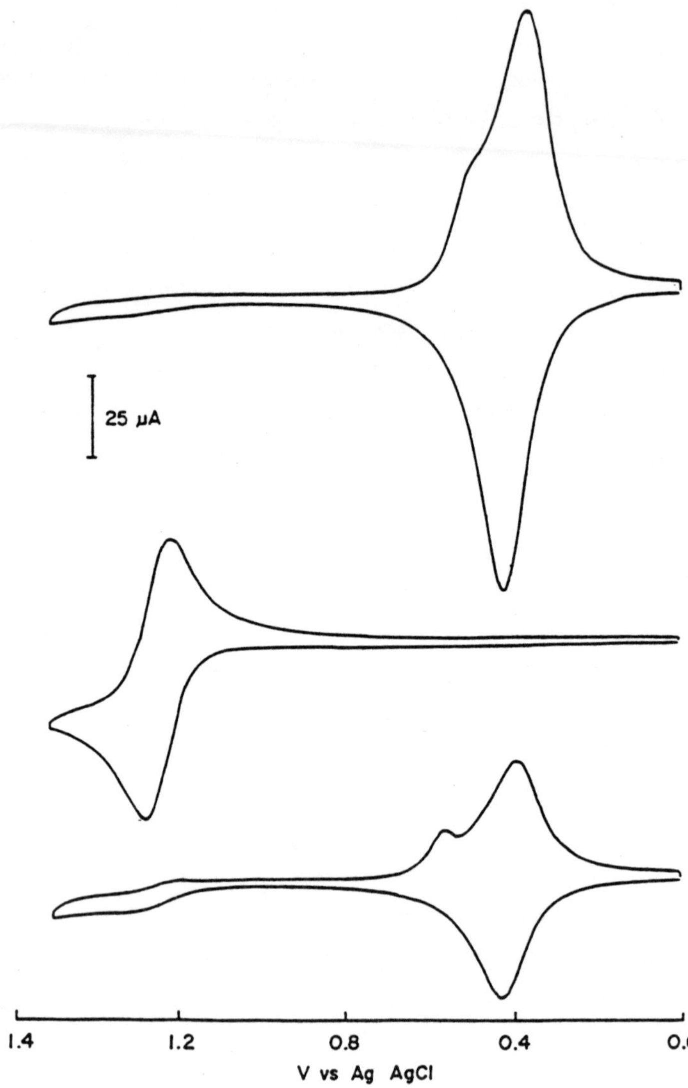

Figure III.7.4: Oxidation of DPA at bare and PPVF film covered platinum electrodes in 0.1 M TBAP/CH_3CN. ν = 50 mV-s^{-1}, [DPA] = 1.0 mM, A = 0.24 cm^2.

Top: oxidation of DPA at 400 Å PPVF film covered electrode

Middle: oxidation of DPA at bare Pt electrode

Bottom: oxidation of DPA at 172 Å PPVF film covered Pt electrode.

reduction in peak current). This demonstrates that even for very thin films, the films are uniform and that any pinholes are necessarily much smaller than 172 Å. For the 400 Å film, the oxidation of DPA is for all intents and purposes totally inhibited. To examine the effect of the size of the substrate, the oxidation of $Ru(Bipy)_3(PF_6)_2$ was examined at the 172 Å film. As expected, even greater inhibition occured for the oxidation of this substrate.

Merz and Bard [3] found that the oxidation of DPA at a PVF film covered Pt electrode ($\Gamma = 1.04 \times 10^{-8}$ moles-cm^{-2}) was unaffected by the film. The same peak currents for DPA oxidation were observed at the PVF film surface as at a bare platinum electrode. The surface coverage of the 172 Å PPVF film was 7.6×10^{-9} moles-cm^{-2}, slightly less than that used by Merz and Bard. This shows that the PPVF film is quite different from PVF, probably more dense and more uniform.

Conclusions

The break-in behavior of PPVF films has been examined as a function of thickness and in different solvent - supporting electrolytes. The break-in behavior for PPVF films made in the second generation reactor was considerably less pronounced than that observed previously for PPVF films made in the prototype reactor. This suggests that PPVF films made in the new reactor are less crosslinked presumably due to the use of lower RF power densities.

The break-in behavior depends markedly on the SSE system. The role of the counterion and the degree of swelling by the solvent are implicated as being very important in the break in process and these are in turn related to the free volume in the PPVF film. It was also shown that the number of ferrocene sites which are electroactive depends on the SSE, implying that in a given SSE system the PPVF film can accommodate a limited amount of counterions and solvent. This further suggests that the PPVF film has a limited swelling capacity in a given SSE. The ferrocene sites which are electroinactive do not seem to participate in the electrochemical behavior of the PPVF films as evidenced by the thermodynamic study in Chapter 8.

Cyclic voltammetric and potential step experiments (current and charge transients) indicate that the charge transport rate changes as the film is oxidized or reduced. In many cases it was seen that the cyclic voltammetric wave shape (e.g. the FWHM) is a consequence of transport kinetics and not thermodynamics. Diffusion coefficients (charge transport) were calculated for films of different thicknesses. However, poor agreement was found for calculated diffusion coefficients for different thickness films. The validity of the model as applied to this system is questioned, since the diffusion coefficient for the charge transport change as a function of the oxidation state of the film.

The importance of evaluation of charge transport rates as determined by linear sweep methods or pulse techniques is dubious as to its utility in aiding in the understanding of electroactive polymer modified electrodes intended for electrocatalytic applications. The charge transport rate measured in transient techniques reflects changes occuring in the polymer (e.g. swelling, incorporation of counterion and solvent) as a consequence of the applied perturbation. However, these charge transport rates may not even closely resemble steady state charge transport in an electroactive polymer film as it would be used in electrocatalysis.

Since these films are not very permeable to dissolved polarizers of sufficient solute size, it would be possible to examine the charge transport rate through the film at steady state. An example of such an experiment is given here for illustrative purposes. A soluble redox couple should be chosen that can easily be oxidized by the PPVF film, but should not readily permeate the film. Using a PPVF film electrode under stirred solution (e.g. rotating disk) at a potential poised so that the film is at least partially oxidized, and that this film will oxidized the redox couple in solution rapidly, then the steady state charage transport rate through the PPVF film can be measured. Varying the thickness of the PPVF film, the potential applied, and the concentration of the redox couple in solution should allow for the unambiguous determination of this charge transport rate. It seems that it is this charge transport rate that would be of most interest and utility for evaluation of the efficiency of electroactive polymers in electrocatalytic applications.

References

1. Schroeder, A., Kaufman, F. B., Patel, V. and Engler, E., J. Electroanal. Chem., 113, 193 (1980).
2. Kaufman, F. B., Schroeder, A. H., Engler, E. M. and Kramer, S. R., J. Amer. Chem. Soc., 102, 483 (1980).
3. Merz, A. and Bard, A. J., J. Amer. Chem. Soc., 100, 3222 (1978).
4. Peerce, P. J. and Bard, A. J. J. Electroanal. Chem., 114, 89 (1980).
5. Chambers, J. Q., J. Electroanal. Chem., in press.

IV. Plasma Volume Synthesis Studies

Studies on organic plasmas have explored the chemistry of cyano compounds and cyanation reactions, the chemistry of COS, and iodination reactions. In addition to mass spectrometric sampling of rf plasmas and emission spectroscopy we have made comparisons with photochemistry to evaluate the neutral, free radical chemistry and comparisons with ion cyclotron resonance spectroscopy to evaluate the ion chemistry.

Our studies of rf cyanation have demonstrated the synthetic and mechanistic importance of these reactions. An example of this kind of process is:

$$CH_2 = CH_2 + C_2N_2 \xrightarrow{rf} CH_2 = CH\text{-}CN$$

We have studied the mechanism of such reactions using physical organic techniques as well as spectroscopy. These studies led to the following mechanism; illustrated for propene.

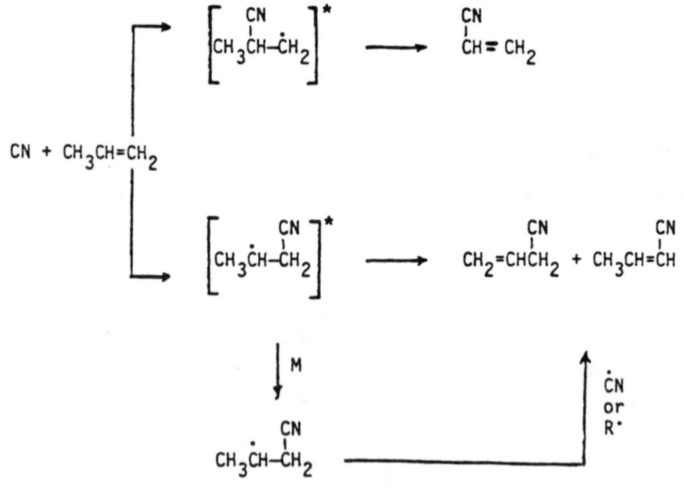

It is clear from our studies that both ground electronic state and excited state CN are produced from C_2N_2 or CH_3CN in the rf plasma. We have measured the emission spectra and detected CN by mass spectral sampling of the plasma zone. This radical, CN, is extremely reactive and accounts for the majority of the stable products whenever these precursors are present.

Careful studies of the emission intensity from CN^*(B state) have elucidated the extent of vibrational and rotational excitation and have documented the dependence of emission intensity on composition (organic additive or argon) pressure and power. These data are simply interpreted. The amount of CN^* emission depends only on the rate of CN^* generation. This latter rate corresponds to electron impact reactions on C_2N_2 or CH_3CN respectively. The rates of electron impact induced fragmentation depends on the electron density and energy distribution - thus power and the pressure of various components.

Perhaps the major undertaking of this project was the construction of mass spectral sampling systems for rf plasmas. We chose to test the capabilities of this system using COS as a reactant. This simple molecule was expected to have relatively simple rf chemistry and we hoped to elucidate the details as well as gain experience with a compound which formed new products.

This work has been published. Carbonyl sulfide was flowed through the plasma zone of an inductively coupled discharge. Mass spectrometry showed the neutrals CO, S, and S_2 as products. The major ions were COS^+, S^+, S_2^+, S_3^+, CO^+, and CS_2^+. The variations in the ionic composition as pressure and power were changed were studied. Reaction products coat the reactor walls after some time and it was shown that this deposit could be sputtered with a CO or Ar plasma to produce sulfur-containing ions. Experiments using a small amount of 2-butyne as coreactant with COS suggested that this hydrocarbon scavenged sulfur atoms which had been formed in the COS reaction. Emission spectroscopy showed major bands due to CO^* and CS^*. It was shown that 5 mol % of SF_6 quenched much of the CS^* emission and it was suggested that CS^* was formed from electron - COS^+ recombination.

In conjunction with this investigation the positive ion-molecule reactions of COS were investigated in an ion cyclotron resonance spectrometer. A variety of reactions in COS/hydrocarbon mixtures have been investigated for C_1-C_4 hydrocarbons which are alkanes, alkenes and alkynes. The formation of organo-sulfur ions was found in reactions in COS/hydrocarbon (C_n) mixtures with $n<4$. Formation of organo-sulfur ions was observed from hydrocarbon ions reacting with COS and COS^+ and S^+ reacting with the hydrocarbons.

Details of the studies are given in Publications 2 through 6 cited in Appendix I.

Appendix I. Publications and Presentations

Publications:

1. M. F. Dautartas, K. R. Mann and J. F. Evans, "Photoassisted Electrocatalytic Reduction of Chloroform and Carbon Tetrachloride Using Plasma Polymerized Vinylferrocene Film Electrodes", J. Electroanal. Chem. 110, 379 (1980).

2. S. Bezuk, Y. H. So and L. L. Miller, "Reactions of Acetonitrile in a Radiofrequency Discharge", J. Org. Chem., 47, 1475 (1982).

3. N.B.H. Henis and L. L. Miller, "Mechanisms of Gas Phase Cyanation of Alkenes and Aromatics". J. Am. Chem. Soc., 105, 2820 (1983).

4. S. J. Bezuk, L. L. Miller and I. Platzner, "Reactions of COS in an RF Plasma", J. Phys. Chem., 87, 131 (1983).

5. L. L. Miller, "Organic Plasma Chemistry", Accounts of Chem. Res., in press. See Appendix III.

6. D. A. Weil, I. Platzner, L. L. Miller and D. A. Dixon, "Positive Ion-molecule Reactions in OCS/Hydrocarbon Mixtures", J. Phys. Chem. submitted.

Papers in Preparation:

1. N.B.H. Henis, and L. L. Miller, "Armoatic Iodination", in preparation.

2. S. J. Bezuk, Ph.D. Thesis, in preparation.

3. M. F. Dautartas and J. F. Evans, "Preparation and Characterization of Plasma Polymerized Polyvinylferrocene Prepared from Neat Vinylferrocene Vapor", to be submitted 7/83 to J. Electrochem. Soc.

4. M. R. Ross, T. R. Hayes, D. M. Ullevig, M. F. Dautartas and J. F. Evnas, "X-ray Photoelectron Spectroscopic and Secondary Ion Mass Spectrometric Thin Film Analysis of Polyvinylferrocene", to be submitted 7/83 to Anal. Chem.

5. E. F. Bowden, M. F. Dautartas and J. F. Evans, "Redox Thermodynamics and Structure of Ferrocene Polymers Film Electrodes", to be submitted 7/83 to J. Electroanal. Chem.

6. D. W. Ernie and H. J. Oskam, "Reliable Detection of Negative Ions from Plasmas", to be submitted to Rev. Sci. Instrument.

Presentations:

1. Gordon Research Conference on Electrochemistry, Santa Barbara, CA: "Charge Transfer Photoelectrochemical REducation of Alky Halides Using Plasma Polymerized Vinylferrocene Electrodes", J. F. Evans, January 29, 1980.

2. Gordon Research Conference on Analytical Chemistry, New Hampton, NH (in session for presentation of ongoing research by young faculty): "Sputtering Yields and Rates for Organic and Organometallic Thin Films, J. F. Evans, August 12, 1980.

3. The Electrochemical Society meeting, Minneapolis, MN: "Catalysis and Transport in Polymer Modified Electrodes", J. F. Evans, March 12, 1981.

4. L. L. Miller, "Organic Plasma Chemistry", Euchem Conference on Organic Electrochemistry, May, 1981.

5. Gordon Conference on Electrochemistry, Ventura, CA; Poster Session: "Super Nernstian Thermodynamics of Plasma Polymerized Vinylferrocene Thin Electrodes", J. F. Evans, January 26, 1982.

6. Middle Atlantic Regional Meeting of the American Chemical Society, Newark, DE: "Characterization of the Thermodynamic and Kinetic Behavior of Plasma Polymerized Polyvinylferrocene Thin Film Electrodes", J. F. Evans, M. F. Dautartas and J. E. McCabe, presented by J. F. Evans, April 23, 1982.

7. Spring Meeting of the Electrochemical Society, Montreal, Quebec, Canada: "Thermodynamic and Kinetic Properties of Plasma Polymerized Vinylferrocene Covered Electrodes", M. F. Dautartas, J. E. McCabe and J. F. Evans, presented by J. F. Evans, May 12, 1982.

8. N.B.H. Henis, "Plasma Cyanation", Gordon Research Conference, 1982.

9. American Vacuum Society Symposium on "Organics in the Future of Surface Science", Bloomington, MN, "Thin Film Analysis of Polymer Layers Using XPS and SIMS", J. F. Evans, May 25, 1983.

10. 17th Great Lakes Regional ACS Meeting (GLRM), St. Paul, MN, "Secondary Ion Mass Spectrometric Thin Film Thickness Measurements of Vinyl Ferrocene Polymers", T. R. Hayes, M. R. Toss, D. M. Ullevig and J. F. Evans, presented by T. R. Hayes, June 3, 1983.

11. 17th Great Lakes Regional ACS Meeting (GLRM), St. Paul, MN, "Measurement of Charge Transfer Impedence in Electroactive Polymer Films", C. C. Chang and J. F. Evans, presented by C. C. Chang, June 3, 1983.

12. 17th Great Lakes Regional ACS Meeting (GLRM), St. Paul, MN, "Redox Thermodynamics and Structure of Ferrocene Polymer Film Electrodes", E. F. Bowden, M. F. Dautartas and J. F. Evans, presented by E. F. Bowden, June 3, 1983.

Appendix II. Personnel

1. Staff: Lorne Chanin, Electrical Engineering
John F. Evans, Chemistry
Larry Miller, Chemistry
Hendrik J. Oskam, Electrical Engineerng
Gary Robinson, Electrical Engineering (until June 1981)

2. Postdoctoral Fellows:

Doug Ernie	(6/31/80-9/15/82)
Neil Henis	(2/30/81-12/15/82)
Thomas Platzner	(6/30/81-5/31/82)
Vidya Jog	(2/30/81-3/15/83)
Edmond Bowden	(12/15/82-6/15/83)

3. Students:

		Degree Awarded
Steve Bezuk	(2/1/80-3/15/82)	PhD, August 1983
Mindaugas Dautartas	(12/15/79-3/15/83)	PhD, January 1982
Dale Ullevig	(9/15/81-12/15/81)	PhD, June, 1983
Dave Swanson	(7/16/80-6/15/82)	
Luis Ferreiro	(12/30/81-7/15/82)	
Julie Brady	(9/15/81-12/15/82)	
Richard Kvitek	(9/15/80-6/15/82)	

Michael Ross: "Secondary Ion Mass Spectrometry of Chemically Bound Organosilane Films and Plasma Polymerized Vinylferrocene Films on Conductive and Semiconductive Substrates", PhD awarded March, 1981. Note: Mike was not supported by this grant as far as salary is concerned, but did work on some aspects of the plasma polymerized vinylferrocene project as noted from title of thesis. His thesis also acknowledge support (supplies, etc.) from ONR grant.

Appendix III. Organic Plasma Chemistry

Larry L. Miller

Department of Chemistry
University of Minnesota
Minneapolis, MN 55455

(Revised)

Because there is an essentially unlimited variety of compositions of matter, it is still possible to achieve interesting and useful results by heating a mixture of compounds and examining the reaction products. In this account the reader will be introduced to a less mundane approach to preparative chemistry, in which a vaporized reactant is passed through an electrical discharge. The reaction zone glows like a neon light and products are formed. These products are sometimes unique and, surprisingly, the method seldom reduces the reactants to molecular rubble.

The glowing region in this reactor is a plasma, a gas which contains some ions and free electrons. Reactions in a plasma are primarily induced by electron impact and, using an RF discharge, it is possible to do electron impact chemistry on a scale of grams. This seemed to be a fascinating, if exotic, approach to organic chemistry and in 1973 my coworker, John Huntington, and I initiated a study of organic plasma chemistry. Our explicit goals were to discover unusual reactions, to exploit the uniqueness of plasmas for preparative purposes and to gain some understanding of the complex processes in plasmas using the methods of physical organic chemistry. This article reports the subsequent studies as well as, related investigations. In the spirit of this journal, some speculation concerning the mechanisms, the utility, and the limitations of plasma chemistry are included.

Plasmas

Since gaseous plasmas and electrical discharges are unfamiliar to many chemists, we begin with a general description.[1] All plasmas are composed of free electrons, ions and various neutral species. The plasma state can be considered the fourth state of matter and although we are more accustomed to solids, liquids, and gases; most of the material in the universe is in the plasma state. In the laboratory we can generate plasmas most easily using electrical energy. Indeed, the chemical effects in an electrical discharge are usually associated with a plasma.

Formation of the plasma is usually initiated by a few free electrons, which are adventitiously present in an ordinary gas. These electrons are accelerated by the applied electrical field, they collide with molecules and are scattered. Occasionally, these collision will impart enough energy to ionize molecules. Ionization, of course, produces more electrons, so the process avalanches until a steady state, determined by the production and scavenging of electrons, is reached. This is the plasma. It is quite conductive and electrical energy can be continuously transmitted to the molecules by the impact of accelerating electrons. In a cool plasma of the type useful for organic chemistry, only a small fraction of the molecules are ionized at any instant. Because reactions are initiated by electron impact, the number of electrons and the distribution of electron energies primarily determine reaction rates. The exact distribution is not known, but it can be crudely conceptualized in terms of a Maxwellian distribution where the mean electron energy is a few eV. This will vary with the composition of the plasma, the pressure, and the applied power. In general, lower pressure and/or higher power will give more energetic plasmas.

Control of the electron energies and density is a key point. The difference between plasma ashing and plasma synthesis is similar to the difference between burning an organic compound and gently heating it. In the discharges of interest here, the neutral molecules have "temperatures" (as measured at the wall) of less than 100°C. On the other hand, the electrons, which are being continuously accelerated, have typical temperatures of 10,000 °K. These fast moving electrons trigger the reactions and in a sense, these glow discharges are like photolyses where photons trigger reactions. In fact, however, a plasma is a unique reaction medium in which the reactants and intermediates find themselves in a sea of fast moving electrons.

Electron impact can lead to vibrationally and electronically excited neutrals and to a variety of ions, any of which could in principle produce products. Wall effects and the effects of excess energy in the products of these gas phase reactions can also have importance. Thus, the dynamics of plasmas which contain organic molecules are extremely complex and there is little mechanistic understanding of the processes involved.

There are many commercial applications of plasma chemistry. A familiar preparative application is the ozonizer, which is an oxygen discharge, run at atmospheric pressure. It is the most efficient method for ozone generation (some 100 x better than photochemistry) on any scale. Indeed it is used for community water purification. Radio-frequency (RF) discharges are also used for large scale processing. In particular the preferred technique for the selective etching of microcircuitry uses an RF discharge. This approach represents a major breakthrough and indicates from a different angle the need for further scientific study of plasmas containing materials more complex than atoms. An understanding of processes in plasmas would also have utility for understanding xerography, prebiotic chemistry, and chemistry in the ionosphere.

Equipment for RF Reactions

Virtually all of the work on organic plasmas has used RF discharges.[2] An inductance coil outside the reaction tube, usually serves to couple the energy from the power supply to the plasma. Although other arrangements are possible, this is convenient and avoids reactions of metallic electrodes with the plasma constituents.

Although there is a belief that reactions occurring in the gaseous plasma are largely frequency independent, RF has proven to be superior to microwave discharges.[2] The gas temperatures are quite high in the latter and some of the material turns to a black, carbonaceous residue. Presumably this is because higher power densities are required to sustain microwave discharges.

The commercial power supply we have used operates at 13.6 MHz and at any power up to 300 W. It is inductively coupled to the chemical system by a coil of copper tubing wound around a 30x2.5 cm glass tube. The organic material is distilled through this tube without a carrier gas at reduced pressure (ca. 0.5 Torr) and products are collected in a cold trap. A plasma can usually be sustained at power values as low as 5W. It is also important to control the flow rate. In a simple apparatus, this can be accomplished by using a teflon needle valve and changing the temperature of the reactant reservoir. Accurate and automatic flow rate and pressure control can be achieved using a commercial flow controller with flow meters, a capacitance manometer and solenoid valves.

In a typical RF experiment, the flow rate is 1 mmol min^{-1}, the residence time in the plasma zone is 50 msec and the power is 50W. These conditions will typically lead to a conversion to products of about 50%. Thus, with the electrical energy input equivalent to that from a light bulb, about 5 mmole of product can be obtained in 10 min. This is truly a preparative scale method and the energy requirements are quite modest.

Reaction Products

In this section several plasma reactions, which are either unusual or instructive are described. The yields in these reactions vary from 40-90%, based on reacted starting material. Four examples, taken from the work of Harold Suhr and coworkers are shown. A more comprehensive review has been published by Professor Suhr.[2]

$$C_6H_5CHO \longrightarrow C_6H_6 + C_6H_5\text{-}C_6H_5 + CO$$

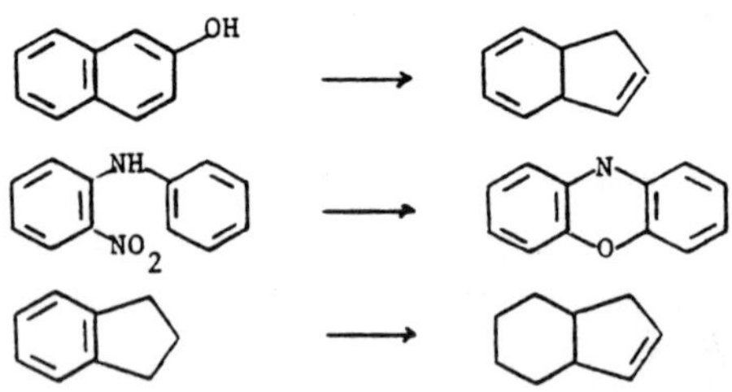

Each of these reactions involves extrusion of a small molecule from the reactant. This is quite common, especially when the small molecule is CO or CO_2. Loss of both CO and CO_2 occurs from anhydrides. For example, starting from either cis- or trans-1,2-dimethylsuccinic anhydride produces the same mixture of cis and trans-2-butene.[3] Since the products are stable, this shows that the reaction is not concerted. The synthesis of a strained product is exemplified by the formation of benzocyclopropene from phthalide.[4]

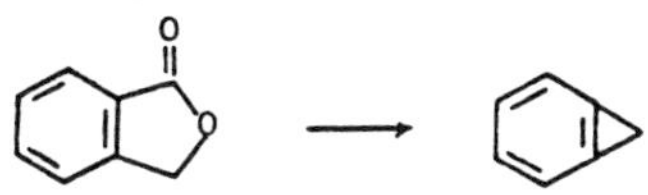

This product mixture contains all of the same by-products found in the pyrolysis of phthalide. A contrast to this complexity is the single product from 3-phenylphthalide.

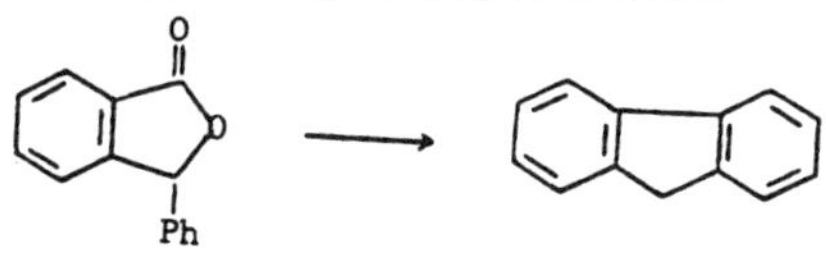

This result is easily rationalized in terms of decarboxylation leading to a diradical, which cyclizes onto the ortho position of the phenyl ring instead of giving the less stable 1-phenylbenzocyclopropene.[4]

Isomerizations are a second general class of RF reactions. Isomerizations of alkenes have been found[5] and some more unusual examples are shown below.[6,7,8]

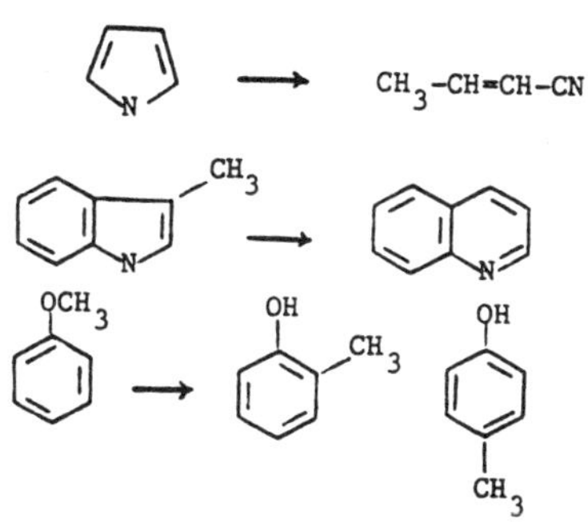

Processes of special synthetic interest are those in which two compounds are simultaneously flowed through an RF apparatus. Only a few experiments of this type are reported in the literature, but they provide promise. One of these involves the plasmolysis of phthallic anhydride in the presence of other compounds.[9] Aniline comes from ammonia as coreactant and phenylacetylene is formed when acetylene was a coreactant. It was proposed that the anhydride lead to benzyne which was then trapped by the additive.

An extremely interesting example of organometallic synthesis has been reported which involves the generation of trifluoromethyl radicals from hexafluoroethane using an RF plasma at 0.5 Torr.[10] Solid metal halides were placed in a Vycor boat at the tail of the plasma region. Trifluoromethyl organometallic compounds resulted. The reaction

$$n(CF_3)_2 + M(X)_n \longrightarrow M(CF_3)_n + \frac{n}{2}X_2$$

M = Te, Hg, Sn, Ge

with HgI_2 produced $Hg(CF_3)_2$ in 95% yield based on HgI_2 consumed. It was shown that this method was especially attractive for thermally unstable products.

The nitrosation of cyclohexane[11] is a final example.

The goal was to produce caprolactam from which nylon 6 can be prepared. Production of the nitroso dimer from cyclohexane was accomplished using a modified ozonizer at 300-760 torr, with a carrier gas. The power was typically 40 W, the conversion rate 1mmol/min and the yield was up to 50% based on either reacted NO or cyclohexane.

These last three examples involving benzyne, metal alkyls and nitrosylation demonstrate that substitution reactions are possible and have some synthetic potential. The examples selected also demonstrate three different approaches to doing selective preparative chemistry in plasmas.

RF Cyanation, A Case Study

Synthetic Aspects

In the past few years, my coworkers and I have studied the RF induced cyanation of unsaturated organic compounds. These studies were triggered by a fortuitous observation. When benzene and acetonitrile were flowed together through the RF reactor, a substantial amount of benzonitrile (1) was formed. Acetonitrile was a somewhat surprising source of cyano groups and so this process was briefly studied.[12] We found that several aromatic compounds could be cyanated in this way, but the yields were not high. Attention was, therefore, turned to the reaction of cyanogen (C_2N_2) with aromatic compounds. This change paid the expected dividends.

When cyanogen was reacted alone in the RF plasma, a somewhat shiny, black solid film resulted. When benzene was reacted alone, a complex mixture of C_2, C_3, C_4 fragments, biphenyl and oligomeric material resulted. In contrast, when a mixture of benzene and cyanogen was reacted, virtually no polymer built up on the reactor wall and considerable yields of benzonitrile (1) resulted.[13]

In this experiment, the two reactants were distilled from separate reservoirs with independently controlled flow rates. A typical reaction required 30 min and in that time 10 g of the aromatic substrate was passed through the apparatus. The inductively coupled reactor described above was used and at 50 W a typical conversion was 50%. In one experiment the yield of 1 based on reacted benzene was 92% and the yield based on reacted cyanogen was 38%. In another experiment with different flow rates the yield was 78% calculated on either basis. It is important to note that neither reagent was present in large excess and both are used efficiently. The energy yield for benzonitrile was typically 1.5 mol $kW^{-1}h^{-1}$.

A variety of substituted benzenes gave aromatic nitriles in each case.[13] No attempt to maximize yield was made, but in several cases more than 60% of the reacted aromatic material was converted to aryl cyanide. The reactions were all relatively unselective, producing not only the ortho (2), meta (3), and para (4) isomers, but also ipso-substituted product (1).

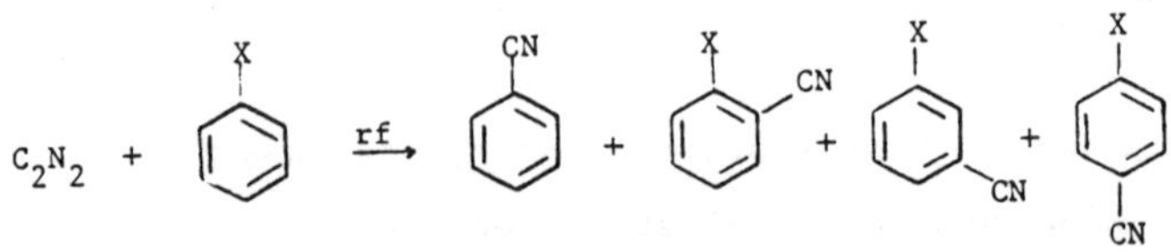

1 2 3 4

$X = CH_3, CF_3, OH, F, Cl, Br, CHO$

With these results in hand, it did not take much ingenuity to initiate a study of alkene and alkyne cyanation. As shown for ethene and 2-butyne, cyanation was successful.[14] Quite reasonable yields of cyano alkenes also resulted from propene, the butenes and from methyl vinyl ketone.

$$CH_2=CH_2 + C_2N_2 \xrightarrow{RF} \underset{\mathbf{5}}{CH_2=\underset{CN}{CH}}$$

$$CH_3C\equiv CCH_3 + C_2N_2 \xrightarrow{RF} CH_3C\equiv C-CN$$

Acrylonitrile **5** is a commercially important product and in this process it can be synthesized from the inexpensive hydrocarbons, ethene or propene. From the latter, the total yield of unsaturated nitriles approached 90%, including acrylonitrile (relative yield 0.70), trans-1-cyanopropene(0.04) cis-1-cyanopropene(0.12), and 3-cyanopropene(0.14).

The energy yields of acrylonitrile from ethene or propene were on the order of 0.5 mol $kW^{-1}h^{-1}$. As discussed below, cyanation can also be achieved photolytically. For comparison, photolysis of 15 Torr of propylene and ICN with a 450 W lamp gave an energy yield of only about 0.005 mol $kW^{-1}h^{-1}$.[15] The Monsanto electrochemical process for producing adiponitrile, where the energy yield is about 2 mol $kW^{-1}h^{-1}$ provides a comparison with a successful industrial process.[16] Thus, although plasma chemistry may involve high energy intermediates, it does not require exorbitant energy expenses. Furthermore, no solvent or catalyst is required.

The cyanation reactions are of some synthetic interest since methods for direct cyanation of aromatics and alkenes are not general. One especially unusual aspect of the chemistry is the prevalence of ipso substitution type reactions, e.g. 1, from C_6H_5X and 1-cyanopropyne from 2-butyne. It is a subject of mechanistic interest to which we will return, but in addition we have recently shown that hydrogen[17] and iodine[18] can be substituted into the ipso position. The iodination is surprising when one considers the low C-I bond strength.

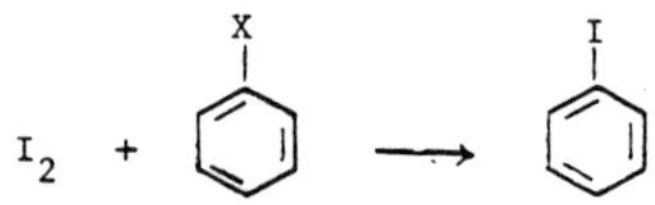

X=H, Br, CHO

Cyanation Mechanisms

A formidable problem. In a plasma with molecular constituents, there are present a large variety of ions, reactive neutrals, and electronically excited species. A further complication is that the electron energy distribution and electron density, which may control reaction rates, are intimately tied to the nature and pressure of the reactant. The reactant is supplying electrons to maintain the plasma; it is also, perhaps by another route, being converted to products. Thus, it is not possible to independently vary the reactant structure or pressure and the electron energy/density. Finally, it is important to know that the plasma zone is not uniform. There are gradients of composition, both along and across the reactor tube. For all these reasons, mechanism elucidation for a molecular plasma cannot have the usual rigor and relative simplicity. Even so, understanding the phenomenon has scientific and technological importance, and it is worthwhile to pursue such studies.

Knowing these complications we have tried to use the methods of physical organic chemistry to elucidate the mechanisms of cyanation. If one considers only the organic chemistry, the following are the minimum steps which are necessary following electron impact initiation.

(a) Cleavage of the NC–CN bond.

(b) Formation of the new C=C–CN bond.

(c) Cleavage of the =C–H (or =C–X) bond.

Therefore, elucidating a mechanistic scheme unelaborated by such effects as excess energy or multiple pathways, simply requires one to arrange the sequence of the steps (a) (b) (c) and specify the precise intermediates (radicals, cations, anions, etc) involved in those steps. It has been proposed that the sequence is actually (a) (b) (c) and that the radical CN, primarily formed by electron impact, is the intermediate which attacks unsaturated organic molecules.[13,14,15]

This species, CN, has been a favorite among spectroscopists. Electronically excited CN radicals ($B^2\Sigma$ and $A^2\pi$) have characteristic emission spectra with many lines, due to vibrational and rotational states. The transition probabilities are high and CN has often been identified in plasmas.[19] The emission spectra of acetonitrile or cyanogen RF plasmas are indeed, completely dominated by CN* emissions[20,21] which give them a characteristic violet glow. It has been shown that the intensity of the CN emission in the RF plasma is controlled by the rate of CN* generation and that the excited state vibrational and rotational energies do not change with reaction conditions. More importantly, the lifetimes of the CN* are short enough, so that virtually all of them decay to ground state CN before collision with the organic substrate. Therefore, there is good evidence for the presence of ground state CN in the plasma. The presence of CN has also been confirmed using mass spectroscopic sampling of these plasmas.[21,22]

Consider now the products from aromatic RF cyanation. The two CN sources, acetonitrile and cyanogen give the same product ratios from toluene. This implies a common intermediate. Most pertinently, the meta/para product ratios, observed when substituted benzenes are reacted, and treated by the Hammett method give a ρ value of -0.5. That value is in concert with expectations for the very reactive, but electrophilic radical, CN, attacking the aromatic compound. It is also quite similar to a ρ=-0.4 found for cyanation achieved by photolyzing ICN in the presence of aromatics in solution.[23]

As indicated above, the plasma reactions give considerable amounts of ipso substitution products. These are not found in the solution phase ICN photolyses. The ipso products have been accomodated by proposing that CN attacks the ipso ring carbon and replaces the substituent X via a vibrationally excited radical adduct (the sigma complex).

The addition of a radical like CN to a double bond is an exothermic reaction and must lead initially to a vibrationally excited radical product.[24] This species can undergo fragmentation reactions before it collides with other molecules, or the wall, and loses this extra energy. At higher pressure there will be a competition between fragmentation and collisional deactivation. Thus, ipso substitution can occur in a low pressure plasma, but not in solution where the activated radical adduct rapidly loses its energy.

The activated radical hypothesis is illustrated for RF cyanation of propene.

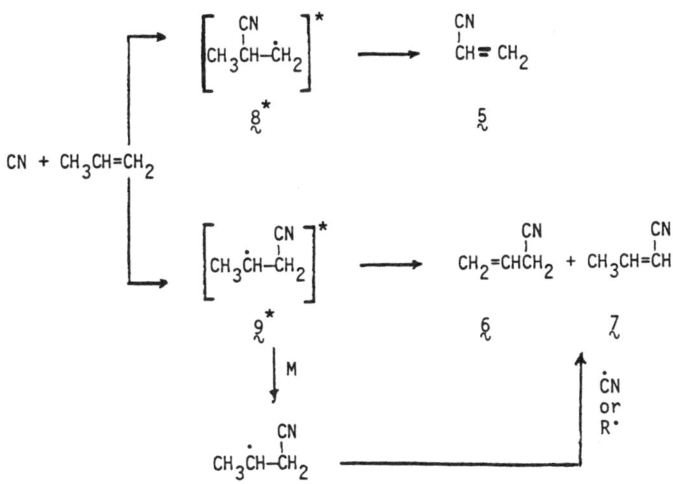

A simple kinetic treatment,[24] using the classical RRK equation can be used calculate the fragmentation rate of **8*** and **9*** and to gain a qualitative understanding of the propene product ratios. As shown in Table I , the loss of methyl from **8*** forming **5** should be extremely rapid and can precede collisional deactivation. This demethylation is in direct competition with hydrogen loss from **8***. The latter is predicted to be much slower and, indeed, 2-cyanopropene is not formed. Radical **9*** can fragment by hydrogen loss, but this will be competitive with collisional deactivation. Products **6** and **7** can, however, also result from bimolecular processes.

Table I. Calculated Rates for Fragmentation of Activated Radicals.

Reaction	$k(sec^{-1})$
$8^* \rightarrow 5$	$3x10^{10}$
$8^* \rightarrow 8$	$1.5x10^{4}$
$9^* \rightarrow 6$	$2x10^{6}$
$9^* \rightarrow 7$	$1x10^{5}$

As a final test, the gas phase photolysis of ICN was performed in the presence of propylene.[15] Previous studies had shown that ICN photolysis at 254nm produced ground state CN.[26] The products and the relative yields from photolysis and plasmolysis were virtually identical and the photolysis yields showed the expected pressure dependence for a hot radical pathway.[15]

This mechanistic description can in fact account for all the product mixtures from unsaturated compounds. The proposed mechanisms have literature precedent in the proposals which have been supported for fluorine atom reactions with alkenes and aromatics.[25]

What about other mechanisms? The involvement of organic radicals, e.g. phenyl, instead of CN, has been considered. Such a mechanism does hold for certain aliphatic cyanation reactions in solution,[27] but does not conform to the data for the plasma reaction. Furthermore, since photolysis of ICN gives the same products as the plasma, it seems that the product forming reactions do not involve ions. Indeed, the products and their relative yields are not compatible with ions as the penultimate, product determining species. It does, however, seem reasonable to propose that ions are involved in the reactions which generate CN. In particular, dissociative electron attachment to cyanogen provides a pathway to CN which is less endothermic than simple dissociation. It is quite likely that CN^- is present in these plasmas.

$$e + C_2N_2 \rightarrow CN + CN^-$$

$$e + C_2N_2 \rightarrow 2CN$$

however, CN^- is not likely to be involved in cyanation, because addition reactions are quite endothermic.

On the other hand, CN^+ is such a high energy species that it is unlikely to be important. Mass spectrometric sampling shows $C_2N_2^+$ and $(C_2N_2)_n^+$ to be the major ions present in the RF plasma of cyanogen.[21] The cation, CN^+, is virtually absent. When the organic is added, hydrocarbon and cyanated hydrocarbon ions are formed.

In summary, we have gained considerable evidence demonstrating that RF cyanation primarily involves ground state CN attacking unsaturated organic compounds to give activated radical adducts. These adducts react rapidly by unimolecular fragmentation or are deactivated and react more slowly by bimolecular hydrogen transfer to give the unsaturated cyano products. Little is known, however, about the initial stages of the process, where the action of electron impact is a central issue.

The Importance of The Electrons

To return to a more general discussion, some further considerations of the importance of electron impact reactions in plasmas and a discussion of the variation in percent conversion and product yield with plasma conditions are important. The data allow some empirical understanding of the limitations of plasma chemistry.

Because plasma reactions are primarily initiated by electron impact the rate of reaction and possibly the mix of products can be controlled by the electron energies and densities. These in turn depend on the power (P) and (inversely) on the pressure (p). In a qualitative sense, increasing P increases the field strength, which increases the acceleration of the electrons. A greater number of electrons with energies above the threshold for causing reaction gives higher conversion of reactant to products. Increasing pressure causes more collisions but fewer of these collisions are above threshold. Since there is more material present the percent conversion is lowered.

Kinetic studies in which the extent of reaction was measured at various values of flow rate, pressure, and power have been reported.[28] The investigations by Bell and coworkers on the reactions of oxygen and carbon dioxide are especially thorough and interesting in that a spacially resolved kinetic model was proposed to explain the results.[28ab] These studies support the proposal that the reaction rate is primarily controlled by the electron energy and density.

A study of cis-2-butene plasmolysis was especially instructive with regard to organic reactions. It was shown[5a] that the extent of conversion depended on power over pressure, Pp^{-1}. Furthermore, there were two competing (not consecutive) reaction pathways, isomerization and fragmentation. At small values of Pp^{-1} only isomerization occurred, at high values only fragmentation. It was, therefore, concluded that the electron energies and densities controlled both the rate and the reaction pathway. More collisions of more energetic electrons with reactant produced higher conversion, but also less selective fragmentation reactions.

The kinetics of the RF plasmolysis of anisole[8] were studied by making a series of runs in which r (flow rate) and P were varied. The data were correlated by a rate law with the form:

$$-\log A/A_o = kPr^{-1} + C$$

where A is the amount of anisole recovered and A_o is the amount passed through the reactor. Although it is rather uninformative mechanistically, this empirically derived rate law is consistent with the idea that electron impact reactions control the rate. Furthermore, this equation has a very interesting implication. That is, that any organic compound will contribute to the control of the electron energies. Experiments with additives explored this point.

The experiments were run using anisole and various coreactants in an apparatus where the flow rate of each reactant could be independently controlled. As additives, nonane, benzaldehyde, tripropylamine, benzonitrile, cumene, benzene, biphenyl, naphthalene, phenylacetylene, aniline, piperylene, carbon tetrachloride, hexafluorobenzene, and anthracene were examined. When the ratio of additive to anisole was <0.1, no effect on the anisole rate of reaction or products could be detected. When the flow rate of additive was increased to be comparable with that of anisole the data demonstrated that the additives act kinetically to cool the plasma in the same way as anisole. Thus, the rate of reaction of anisole in these mixtures could be correlated with the total flow rate of organic. It did not correlate with the r of anisole. Since under these conditions r is approximatly proportional to p this result is consistent with the idea that electron energies and densities are controlled by the total pressure and are not very sensitive to the structure of the organic compounds.

The variable r can be considered to contain contributions from both the pressure and residence time in the plasma zone (τ). Thus, it is of interest to independently vary these parameters. This is not easy to do in a quantitative fashion because the size of the plasma zone (which affects τ) often changes with pressure. We have used a pulsed RF discharge to avoid this problem and to directly study the dependence of reaction rate on τ. In this experiment, the on/off time ratio at constant P, p effectively changed the amount of time the compound was in the plasma zone. An appropriate duty cycle was in the range of a few milliseconds. The kinetic data so obtained for the conversion of CCl_4 to C_2Cl_6 and C_2Cl_4 revealed that the log of the extent of conversion was proportional to τ, P and p^{-1}.[29]

Finally, we return to C_2N_2 plasmas to further illustrate the connection between inelastic electron scattering and reaction rates. At low pressure, the rate of emission from the electronically excited CN (^1B) is substantially greater than the collision rate. Therefore, the steady state intensity of the emission[19] is a measure of the rate of B state generation. It was found that the intensity was approximately proportional to P. At constant power and constant total pressure, the mole fraction of C_2N_2 was varied by dilution with argon or organic compounds. As expected, addition of the organic led to rapid diminution in the amount of CN generated. Addition of argon, however, had little effect on the emission and there was no emission from argon. Argon, with only high energy electronic states and a high ionization potential, does not scatter electrons as effectively as C_2N_2 or organic compounds. As the C_2N_2 pressure decreases, the number of electrons above threshold increases, and this balances the decrease in the number of C_2N_2 molecules available for reaction.

A variety of other data have been collected including relative rate data for several organic compounds. All the data available suggest that electron impact events primarily control reactivity and these events are not very selective. It should be realized that in a quantitative way each rate is the result of a complex variety of chemical processes. Not only are there different overall chemical reactions involved, but there must often be pathways for regenerating reactant, e.g., by radical recombination or excited state quenching, and for indirectly consuming reactant in non-electron impact events including photolysis from emitted light. These will obviously contribute to the measured rate to different extents for each compound. The important conclusion, however, is that conversion rates depend strongly on the electron energy/density.

Conclusion

The studies cited indicate the potentialities and limitations of plasma chemistry. It is a useful method for activating normally inert compounds and it could have utility as an energy efficient method, not requiring solvent or catalyst. However, plasmas will not be sufficiently selective for the synthesis of complex molecules.

Mechanistically, reaction rates are primarily controlled by the electron energy/density. The isolated reaction products can usually be explained by neutral radical or excited state precursors and consideration of the structure and internal energy of these precursors can give mechanistic insight. The mechanisms by which these are generated has not been elucidated and could in part involve ions. On the other hand, although a plasma is an ionized gas there is virtually no reliable evidence that the major products isolated come from ionic reactions. Such evidence will be difficult to obtain for cationic reactions, but the possiblity of doing understandable and useful preparative anionic chemistry with the low energy electrons in plasmas is quite worthy of investigation.

Molecular plasma chemistry has not received systematic study and most of the available information refers to a particular style of RF reactor. This is obviously not the only way to perform these reactions and future explorations should hold many surprises.

Acknowledgement

My coworkers Drs. Huntington, Szabo, Tezuka, Schmalzl, So, Tokuda, Henis and Mr. Bezuk are acknowledged. Because this is an uncommon area of research, our intramural interactions have been especially important. The initial encouragement of Professor Harold Suhr was also important. Support from the National Science Foundation, and the Office of Naval Research with seed money from the Research Corporation and the A.P. Sloan Foundation are gratefully acknowledged.

References

1. Fundamentals of molecular plasma chemistry are discussed in: (a) Hollahan, J. R.; Bell, A. T., Eds. "Techniques and Applications of Plasma Chemistry"; Wiley: New York, 1974. (b) Capitelli, M.; Molinari, E. Top. Curr. Chem. 1980, 90,59. (c) Chapman, B. "Glow Discharge Processes", Wiley: New York, 1980 (d) Shen, M.; Bell, A. T., Eds., "Plasma and Polymerization"; American Chemical Society: Washington, DC, 1979.

2. (a)H. Suhr in "Techniques and Applications of Plasma Chemistry", J. R. Hollahan and A. T. Bell, Ed., Wiley, New York, N.Y., 1974, Chap. 2. (b) H. Suhr, Angew. Chem., Int. Ed. Engl., 11, 781 (1972).

3. Unpublished work of J. Huntington and L. L. Miller.

4. M. Tokuda, L. L. Miller, A. Szabo and H. Suhr, J. Org. Chem. 44, 4504 (1979).

5. (a) J. G. Huntington and L. L. Miller, J. Am. Chem. Soc., 98, 8101 (1976); (b) H. Suhr and U. Schucker, Synthesis, 431 (1970).

6. H. Suhr and U. Schoch, Chem. Ber. 108 3469 (1975).

7. H. Suhr and R. I. Weiss, Liebigs Ann. Chem., 760, 127 (1972).

8. M. Tezuka and L. L. Miller, J. Am. Chem. Soc., 100, 4201 (1978).

9. H. Suhr and A. Szabo, Liebigs Ann. Chemie, 752 37 (1971).

10. J. A. Morrison and R. J. Lagow, Inorg. Chem., 16, 1823 (1977). Lagow, et. at J. Am. Chem. Soc. 3229 (1978).

11. J. Wagenknecht, I. and E. C. Prod. Res. and Develop., 10, 184 (1971).

12. L. L. Miller and A. Szabo, J. Org. Chem., 44, 1670 (1979).

13. N.B.H. Henis, Y. H. So, L. L. Miller, J. Am. Chem. Soc., 103, (1981).

14. N.B.H. Henis and L. L. Miller, J. Am. Chem. Soc., 104, 2526, (1982).

15. N.B.H. Henis and L. L. Miller, submitted for publication.

16. M. Baizer, Ed., "Organic Electrochemistry", M. Dekker, New York, Chap. 28, 1973.

17. P. Schmalzl and L. L. Miller, Plasma Chem. and Plasma Processing, 2, 43 (1982).

18. Unpublished results of N.B.H. Henis and L. L. Miller.

19. J. E. Nicholas and C. A. Amadio, J. Chem. Soc., Faraday Trans. 1, 4201, 1978. M. R. Gorbal and M. I. Savadatti, Chem. Rev. 82, 527 (1982).

20. Y. H. So, S. Bezuk, and L. L. Miller, J. Org. Chem. 47, 1475, (1982).

21. S. Bezuk, L. L. Miller, unpublished result.

22. Although mass spectroscopic sampling of the active plasma seems a natural method to assist in mechanism elucidation, it is a method fraught with difficulty. This difficulty arises because the plasma must be sampled through a pin-hole in a wall and for several reasons, the material passing through the pin-hole will not accurately represent the bulk of the plasma zone.

23. P. Spagnolo, L. Testaferri, M. Tiecco, J. Chem. Soc B, 2006, (1971).

24. P. J. Robinson and K. A. Holbrook, "Unimolecular Reactions", Wiley-Interscience, New York, 1972.

25. (a) J. A. Cramer and F. S. Rowland, J. Am. Chem. Soc. 96 6579 (1974); (b) K. Shobatake, Y. T. Lee, and S. A. Rice, J. Chem. Phys. 59, 1435, (1973).

26. M. J. Sabety-Dzvonik and R. J. Cody, J. Chem. Phys. 66, 125, (1977).

27. D. D. Tanner and J. J. Bunce, J. Am. Chem. Soc., 91, 3028, (1969).

28. (a) H. Kobaysahi, M. Shen and A. T. Bell, J. Macromol. Sci. Chem., A8(2), 373 (1974), (b) L. C. Brown and A. T. Bell, Ind. Eng. Fund., 13, 203 (1979), (c) G. I. Bebeshko and E. N. Eremin, Russ., J. Phys. Chem., 48, 49 (1974), (d) G. Nolet, J. Electrochem. Soc., 103,4 (1978).

29. Unpublished work of M. Tezuka and L. L. Miller.

Other Noyes Publications

DEPOSITION TECHNOLOGIES FOR FILMS AND COATINGS Developments and Applications

by

Rointan F. Bunshah
John M. Blocher, Jr.
Thomas D. Bonifield
John G. Fish
P.B. Ghate
Birgit E. Jacobson
Donald M. Mattox
Gary E. McGuire
Morton Schwartz
John A. Thornton
Robert C. Tucker, Jr.

This book presents a unique collection of current knowledge on deposition technologies up to and including the most recent state of the art. It breaks new ground in the extensive coverage for those processes used in high technology, covering the entire spectrum from thin films to bulk coatings.

The contents deal with various technologies for the deposition of films and coatings, and the resulting microstructure, properties and applications. It covers the subjects of Evaporation, Ion Plating, Sputtering, Chemical Vapor Deposition, Electrodeposition from aqueous solution, Plasma and Detonation Gun Coating techniques and Polymeric Coatings. Additionally several other subjects common to the above technologies are also covered. They include: Adhesion of Coatings, Cleaning of Substrates, role of Plasmas in Deposition Processes, Characterization of Thin Films, and the Application of Deposition Technologies in Microelectronics.

Written by leading researchers in the field, this book will be useful and needed by scientists, engineers and managers working in industries associated with coatings for optical, electrical, mechanical, chemical and decorative applications. A **complete listing of chapter titles** is given below.

CONTENTS

ISBN 0-8155-0906-5 (1982) **585 pages**